Discourse on Method

and

Meditations

on

First Philosophy

Discourse on Method

and

Meditations
on
First Philosophy

René Descartes

Translated by
DONALD A. CRESS

Hackett Publishing Company

René Descartes: 1596–1650

DISCOURSE ON METHOD was originally
   published in 1637
MEDITATIONS ON FIRST PHILOSOPHY was
   originally published in 1641

Copyright © 1980 by Hackett Publishing Company, Inc.
All rights reserved
Printed in the United States of America
SECOND EDITION *Fifth Printing* 1986

Cover design by Richard L. Listenberger
Interior design by James N. Rogers

For further information please address the publisher,
P.O. Box 44937, Indianapolis, Indiana 46204

Library of Congress Cataloging in Publication Data

Descartes, René, 1596–1650.
   Discourse on method and Meditations on first
philosophy.

   Translations of Discours de la méthode and
Meditationes de prima philosophia.
   Bibliography: p.
   1. Science—Methodology.   2. First philosophy.
I. Descartes, René, 1596–1650. Meditationes de
prima philosophia. English.   II. Cress, Donald A.
III. Title.
B1848.E5C73 1980b   194   80–10809
ISBN 0–915144–85–9
ISBN 0–915144–84–0 pbk.

Contents

DISCOURSE ON METHOD

MEDITATIONS ON FIRST PHILOSOPHY

EDITOR'S PREFACE

René Descartes was born March 31, 1596, in a small town in Touraine called La Haye (now called La Haye–Descartes or simply Descartes). When he was about ten years old, his father sent him to the Collège Henri IV at La Flèche, a newly formed school which was soon to become the showcase of Jesuit education and one of the outstanding centers for academic training in Europe. Later in his life Descartes looked with pride on the classical education he received from the Jesuits, even though he did not always find what the Jesuits taught him agreeable. He especially found the scholastic Aristotelianism taught there distasteful, although he did cherish his training in many other disciplines—particularly mathematics.

Descartes left La Flèche in 1614 to study civil and canon law at Poitiers, and by 1616 had received the baccalaureate and licentiate degrees in law. In 1618 Descartes joined the army of Prince Maurice of Nassau as an unpaid volunteer, but apparently he never saw combat. He seems to have been more interested in using military service as a means of seeing the world.

During a tour of duty in Germany, events of lifelong importance happened to Descartes. In November of 1619 he was sitting in a *poêle*, a small stove–heated room, meditating on the disunity and uncertainty of his knowledge. He marveled at mathematics, a science in which he found certainty, necessity, and precision. How could he find a basis for all knowledge so that it might have the same unity and certainty as mathematics? Then, in a blinding flash, Descartes saw the method to be pursued for putting all the sciences, all knowledge, on a firm footing. This method made clear both how new knowledge was to be achieved and how all previous knowledge could be united and integrated. That evening Descartes had a series of dreams that seemed to put a divine stamp of approval on his project. Shortly thereafter Descartes left military service.

Throughout the early part of his life, Descartes was plagued by a sense of impotence and frustration about the task he had set about to accomplish: a new and stable basis for all knowledge. He had the programmatic vision, but he seemed to despair of being able to work it out in detail. Thus, perhaps we have an explanation for the fact that Descartes, during much of the 1620s, threw himself into the pursuit of the good life. Travel, gambling, and dueling seemed especially to attract his attention.

vii

This way of life ended in 1628, when, through the encouragement of Cardinal de Bérulle, Descartes decided to see his program through to completion. He left France to avoid the glamour and the social life; he renounced the distractions in which he could easily lose himself and forget what he knew to be his true calling. He departed for Holland, where he would live for the next twenty years.

It was during this period that Descartes began his *Rules for the Direction of the Mind* and wrote a short treatise on metaphysics, although the former was not published during his life time and the latter seems to have been destroyed by him. Much of the early 1630s was taken up with scientific questions. However, Descartes' publication plans were abruptly altered when he learned of the trial of Galileo in Rome. Descartes decided, as Aristotle had centuries before, that philosophy would not be sinned against twice. He suppressed his scientific treatise, *The World or Treatise on Light*.

In 1637 Descartes published in French a *Discourse on the Method for Conducting One's Reason Rightly and for Searching for Truth in the Sciences;* it introduced three treatises which were to exemplify the new method; one on optics, one on geometry, and one on meteorology. Part IV of the introductory *Discourse* contained, in somewhat sketchy form, much of the philosophical basis for constructing the new system of knowledge.

In response to queries about this section, Descartes prepared a much lengthier discussion of the philosophical underpinnings for his vision of a unified and certain body of human knowledge. This response was to be his *Meditations on First Philosophy*, completed in the spring of 1640— but not published until August, 1641. Attached to the *Meditations* were sets of objections and queries sent by readers who had read the manuscript plus Descartes' replies to each set.

The period following the publication of the *Meditations* was marked by controversy and polemics. Aristotelians, both Catholic and Protestant, were outraged; many who did not understand Descartes' teachings took him to be an atheist and a libertine. In spite of all of this clamor, Descartes hoped that his teachings would replace those of Aristotle. To this end he published in 1644 his *Principles of Philosophy*, a four–part treatise which he hoped would supplant the Aristotelian scholastic manuals used in most universities. The last important work to be published during his lifetime was his *Passions of the Soul*, in which Descartes explored such topics as the relationship of the soul to the body, the nature of emotion, and the role of the will in controlling the emotions.

In 1649 Queen Christina of Sweden convinced Descartes that he should come to Stockholm in order to teach her philosophy. Christina seems to have regarded Descartes more as a court ornament for her amusement and edification than as a serious philosopher; however, it was the brutal winter of 1649 that proved to be Descartes' undoing. Of the climate in Sweden Descartes was to say: "it seems to me that men's thoughts freeze here during winter, just as does the water." Descartes caught pneumonia early in February of 1650, and, after more than a week of suffering, died on February 11.

SELECTED BIBLIOGRAPHY

A. USEFUL BIBLIOGRAPHIES

Sebba, Gregory. *Bibliographia Cartesiana: A Critical Guide to the Descartes Literature 1800–1960.* The Hague: Nijhoff, 1964. This is the basic bibliographical tool of Descartes scholarship. It contains a large number of annotations and cross-references; it is well indexed by person and by subject matter. Although it is rather weak in its coverage of twentieth-century analytical literature on Descartes, this defect is rectified somewhat by Willis Doney's bibliography in his *Descartes: A Collection of Critical Essays* (New York: Doubleday, 1967), pp. 369–386. This latter bibliography is divided by subject matter and is concerned principally with English titles.

Doney, Willis. "Some Recent Work on Descartes: A Bibliography," *Philosophical Research Archives*, 2, pp. 545–567. This covers 1966–1976, picking up from Doney's earlier bibliography. Divided well by subject matter; not annotated, but still very useful.

Equipe Descartes (C.N.R.S., Paris), "Bulletin Cartesien," published annually since 1972 in *Archives de Philosophie.* These annual "bulletins" are an excellent international bibliography of recent Cartesian studies; annotations are almost exclusively in French.

B. STANDARD EDITION

Adam, Charles and Paul Tannery, eds. *Oeuvres de Descartes.* 13 vols. Paris: Cerf, 1897–1913. (Volumes 1–11 contain Descartes' writings; vol. 12 contains Charles Adam's *Vie et Oeuvres de Descartes;* vol. 13 is a supplementary volume containing correspondence, biographical material, and various indices.) This edition has been recently updated (Paris: Vrin, 1964–) and additional correspondence has been attached to the earlier edition at the back of various volumes. More accurate identifications of dates and addresses have been supplied.

C. ENGLISH TRANSLATIONS

Cottingham, John, tr. *Descartes' Conversation with Burman*. Oxford: Oxford University Press, 1976. Housed in the Library of the University of Göttingen is a manuscript that purports to chronicle a discussion between Descartes and the young Dutch theologian Francis Burman. Burman had chosen several texts from Descartes' writings for discussion. Sometimes he would criticize the doctrine in the text; sometimes he would simply ask for clarification. Descartes' replies are always interesting and nearly always shed new light on difficult passages in his printed works.

Haldane, Elizabeth S. and G. R. T. Ross. *The Philosophical Works of Descartes*. 2 vols. 2nd ed., corr. Cambridge, England: Cambridge University Press, 1931. For many years this often reprinted set of volumes has been the basic translation of many of Descartes' central works. Although not the best translation, it is still the most complete English translation of Descartes' philosophical writings.

Hall, Thomas Steele. *Treatise of Man*. Cambridge, Mass., Harvard University Press, 1972. Descartes' French text plus translation and commentary.

Kenny, Anthony, ed. and tr. *Philosophical Letters*. Minneapolis: University of Minnesota Press, 1981. A valuable source of Descartes' ideas. It complements the written works now available in English translation and should have a corrective influence on scholarship totally dependent on the Haldane and Ross edition.

Mahoney, Michael Sean, tr. *Le Monde ou Traité de la lumière*. New York: Abaris Books, 1979.

Miller, Reese P. and Valentine R. Miller, trs. *Principles of Philosophy*. Hingham, Mass.: Kluwer Academic, 1983.

Olscamp, Paul J. tr. *Discourse on Method, Optics, Geometry, and Meteorology*. Indianapolis: Bobbs-Merrill, 1965.

Robert, Walter. tr. *Compendium of Music*. Rome: American Institute of Musicology, 1961.

D. BOOKS ON DESCARTES

Beck, Leslie. *The Metaphysics of Descartes*. Oxford: Oxford University Press, 1965.

Butler, R. J. *Cartesian Studies*. Oxford: Basil Blackwell, 1972.

Caton, Hiram. *The Origin of Subjectivity: An Essay on Descartes.* New Haven: Yale University Press, 1973.

Curley, E. M. *Descartes Against the Sceptics:* Cambridge, Mass.: Harvard University Press, 1978.

Frankfurt, Harry G. *Demons, Dreamers, and Madmen: The Defense of Reason in Descartes' Meditations.* Indianapolis: Bobbs-Merrill, 1970.

Gibson, A. Boyce, *The Philosophy of Descartes.* London: Methuen and Co., 1932.

Gilson, Etienne, *Discours de la méthode: texte et commentaire.* Paris: Vrin, 1976 (5th ed).

Hooker, Michael. *Descartes, Critical and Interpretive Essays.* Baltimore: Johns Hopkins University Press, 1978.

Kenny, Anthony. *Descartes, A Study of his Philosophy.* New York: Random House, 1968.

Sesonske, Alexander, and Noel Fleming, eds. *Meta-Meditations: Studies in Descartes.* Belmont: Wadsworth Publishing Co., 1965.

Williams, Bernard. *Descartes, the Project of Pure Enquiry.* London: Penguin Books, 1978.

Wilson, Margaret. *Descartes.* London: Routledge and Kegan Paul, 1978.

Discourse
on
the Method for
Rightly Conducting
One's Reason
and
for Seeking Truth in
the Sciences

NOTE ON THE TRANSLATION

The translation is based on the original French version (1637) of the *Discourse on Method* found in volume six of the Adam and Tannery edition of Descartes' works (Paris: Vrin, 1965). The numbers in the margins of this translation refer to the pagination of the Adam and Tannery edition.

D.A.C.

The editor and publisher acknowledge with appreciation the help of Thérèse-Anne Druart in emending this edition of the *Discourse on Method*

DISCOURSE ON THE METHOD FOR RIGHTLY CONDUCTING ONE'S REASON AND FOR SEEKING TRUTH IN THE SCIENCES

If this discourse seems too long to be read at one sitting, one might split it into six parts. In the first, one will find various discussions concerning the sciences. In the second part, the chief rules of the method which the author has been seeking. In the third part, some of the rules of morality which the author has derived from this method. In the fourth part, the reasons by which the author proves the existence of God and of the human soul, which are the foundations of his metaphysics. In the fifth part, the order of the questions in physics which the author has sought—and particularly the explanation of the heart's movement and other difficulties which pertain to medicine, as well as the difference between our soul and that of animals. And in the final part, what things the author believes are required to advance further in the study of nature than the author has done, and what reasons moved him to write.

PART ONE

Good sense is the most evenly distributed commodity in the world, for each of us considers himself to be so well endowed therewith that even those who are the most difficult to please in all other matters are not wont to desire more of it than they have. It is not likely that anyone is mistaken about this fact. Rather, it provides evidence that the power of judging rightly and of distinguishing the true from the false (which, properly speaking, is what people call good sense or reason) is naturally equal in all men. Thus the diversity of our opinions does not arise from the fact that some people are more reasonable than others, but simply from the fact that we conduct our thoughts along different lines and do not consider the same things. For it is not enough to have a good mind; the main thing is to use it well. The greatest souls are capable of the greatest vices as well as of the greatest virtues. And if they always follow the correct path, those who move forward only very slowly can make much greater progress than do those who run and stray from it.

For myself, I have never presumed that my mind was in any respect more perfect than anyone else's. In fact, I have often longed to have as quick a wit or as precise and distinct an imagination or as full and responsive a memory as certain other people. And I know of no other qualities that aid in the perfection of the mind. For as to reason or good sense, given that it alone makes us men and distinguishes us from animals, I pre-

1

fer to believe that it exists whole and entire in each one of us. In this belief I am following the standard opinion held by philosophers who say that there are differences of degree only among accidents, but not among forms or natures of individuals of the same species.

But I shall have no fear of declaring that I think I have been fortunate; I have, since my youth, found myself on paths that have led me to certain considerations and maxims from which I have formed a method by means of which, it seems to me, I have the ways to increase my knowledge by degrees and to raise it gradually to the highest point to which the mediocrity of my mind and the short span of my life can allow it to attain. For I have already reaped from it such a harvest that, though as regards judgments I make of myself, I try always to lean toward caution, rather than toward presumption, and though, looking with a philosopher's eye at the various actions and enterprises of men, there is hardly one that does not seem to me vain and useless, I always take immense satisfaction in the progress that I think I have made in the search for truth; and I envisage such hopes for the future that if, among the occupations of men, as men, there is one which may be solidly good and important, I dare believe that it is the occupation I have chosen.

All the same, it could be that I am mistaken; and what I have taken for gold and diamonds may perhaps be nothing but copper and glass. I know how much we are prone to be mistaken in those things that deeply affect us, and also how judgments made by our friends must be held suspect when these judgments are in our favor. But I would be very happy to show in this discourse the paths that I have followed and to present my life as if in a picture, so that each person may judge it; learning what people commonly think about it may be a new means of teaching myself, which I shall add to those that I am accustomed to employing.

Thus my purpose here is not to teach the method that everyone ought to follow in order to conduct his reason correctly, but merely to show how I have tried to conduct mine. Those who take it upon themselves to give precepts ought to regard themselves as more competent than those to whom they give them; and if they are found wanting in the least detail, they are blameworthy. But, putting forward this essay as merely a history—or, if you prefer, a fable—in which, among the examples one can imitate, one also finds perhaps several others which one is right in not following, I hope that the essay will be useful to some, while harmful to none, and that my openness will be to everyone's liking.

I have been raised on letters from my childhood, and because I was convinced that through them one might acquire a clear and steady knowledge of everything that is useful for life, I possessed a tremendous desire to learn them. But, as soon as I completed this entire course of study, at the end of which one is ordinarily received into the ranks of the learned, I changed my mind entirely. For I was embarrassed by so many doubts and errors, which appeared in no way to profit me in my attempt at learning, except that more and more I discovered my ignorance. And nevertheless, I was in one of the most celebrated schools of Europe, where I thought there ought to be learned men—if in fact there were any such men in the world. I learned everything the others learned; and, not judging the disciplines taught there to be enough, I even went through every book I could lay my hands on that treated those disciplines considered the most curious and unusual. Moreover, I knew what judgments others were making about me; and it was apparent to me that I was rated no less than my peers, even though there already were those among them who were destined to succeed our teachers. And finally our century seems to me just as flourishing and as fertile in good minds as any of the preceding ones. All of this caused me to feel free to judge everybody else by myself, and to think that there has been no body of knowledge in the world which was of the sort that I had previously hoped to find.

Yet I never ceased admiring the academic exercises with which we occupied ourselves in school. I realized that the languages one learns there are necessary for the understanding of classical texts; that the gracefulness of fables awakens the mind; memorable deeds sustain it; and, read with discretion, they aid in forming one's judgment; that reading good books is like a conversation with the noblest people of past centuries—their authors—indeed, even a studied conversation in which they discover only the best of their thoughts; that eloquence has incomparable power and beauty; that poetry has a ravishing delicacy and sweetness; that mathematics contains very subtle inventions that can serve as much to satisfy the curious as to faciltate the arts and to diminish men's labor; that writings dealing with morals contain many lessons and exhortations to virtue that are quite useful; that theology teaches one how to go to heaven; that philosophy provides the means of speaking with probability about all things and of being held in admiration by the less learned; that law, medicine, and the other sciences bestow honors and riches upon those who cultivate them. And thus it is good to have examined all of these disciplines, even the most superstition-ridden and false of them, so that one might know their true worth and guard against being deceived.

But I thought that I had already given enough time to languages and also even to the reading of ancient books—to their histories and to their fables. For it is about the same to converse with those of other centuries as it is to travel. It is good to know something of the customs of various peoples, in order to judge our own more soundly and not to think that everything that is contrary to our way of doing things is worthy of scorn and against reason, as those who have seen nothing commonly think.

But when one takes too much time traveling, one becomes finally a stranger in one's own country; and when one is too curious about things that took place in past centuries, one ordinarily remains quite ignorant of what is taking place in ones own country. Moreover, fables make one imagine many events to be possible which really are impossible. And even the most accurate histories, if they neither alter nor augment the significance of things, in order to render them more worthy of being read, at least almost always omit the basest and least illustrious details, and thus the remainder does not appear as it really is, and those who govern their own conduct on the basis of the examples drawn from it are subject to falling into the extravagances of the knights of our novels and to conceiving plans that are beyond their powers.

7

I held eloquence in high regard and I loved poetry, but I believed that they were both gifts of the mind—not fruits of study. Those who possess the most forceful power of reasoning and who best order their thoughts so as to render them clear and intelligible can always best persuade one of what they are proposing, even if they speak only the dialect of Lower Brittany and have never learned rhetoric.[1] And those who are in possession of the most pleasing rhetorical devices and who know how to express them with the greatest of embellishment and sweetness will not fail to be the greatest poets, even if the art of poetry be unknown to them.

I took especially great pleasure in mathematics because of the certainty and the evidence of its arguments. But I did not yet notice its true usefulness and, thinking that it seemed useful only to the mechanical arts, I was astonished that, because its foundations were so solid and firm, no one had built anything more noble upon them. On the other hand, I compared the writings of the ancient pagans who discuss morals to very proud and magnificent palaces that are built on nothing but sand and mud. They place virtues on a high plateau and make them appear to be valued more than anything else in the world, but they do not sufficiently instruct us about how to know them; and often what they call by such a fine-sounding name is nothing more than insensibility, pride, despair, or parricide.

8

I revered our theology, and I desired as much as the next man to go to heaven; but having learned as something very certain that the road is no less open to the most ignorant than to the most learned, and that the revealed truths leading to it are beyond our understanding, I would not have dared to subject them to my feeble reasonings. And I believed that, in order to undertake the examination of these truths and to succeed in doing so, it was necessary to have some extraordinary assistance from heaven and to be more than a man.

---

[1] This dialect was considered rather barbarous and hardly suitable for sophisticated literary endeavors.

Of philosophy I shall say only that, aware that philosophy has been cultivated over several centuries by the most excellent minds who have ever lived and that, nevertheless, there is nothing about which there is not some dispute—and thus nothing that is not doubtful—I was not so presumptuous as to hope to fare any better than the others; and that, considering how there can be various opinions that are held by learned people about the very same matter without there ever being any more than one opinion being true, I took to be virtually false everything that was merely probable.

As to the other sciences, since they derive their principles from philosophy, I judged that one could not have built anything solid upon foundations having so little firmness. And neither the honor nor the monetary gain they promised were sufficient to incite me to learn them, for I did not see myself, thank God, as being in a condition that forced me to make a trade out of knowledge for the enhancement of my fortune. And although I did not make a point of rejecting glory in the manner of the cynic, still I made light of that glory that was acquired only through false pretenses. And finally, as to the evil doctrines, I believed I already knew them for what they were worth, so as not to be subject to being deceived either by the promises of an alchemist, by the predictions of an astrologer, by the importures of a magician or by the artifices or boasting of anybody who makes a point of claiming to have more knowledge than he actually has.

That is why, as soon as age permitted me to escape the tutelage of my teachers, I left the study of letters completely. And resolving to search for no other knowledge than what I could find within myself, or in the great book of the world, I spent the rest of my youth traveling, seeing various courts and armies, frequenting peoples of varied humors and conditions, gathering varied experiences, testing myself in the encounters which fortune sent my way, and everywhere so reflecting upon what came my way that I could draw some profit from it. For it seemed to me that I could discover much more truth in the reasonings that each person makes concerning matters that are important to him, whose outcome ought to cost him dearly later on if he has judged incorrectly, than in those reasonings that a man of letters makes in his private room, which touch on speculations producing no effect, and which for him have no other consequence except perhaps that the more they are removed from common sense, he will derive all the more vanity from them, for he will have to employ that much more wit and artifice in attempting to make them probable. And I have always had an especially great craving for learning to distinguish the true from the false, to see my way clearly in my actions, and to go forward with confidence in this life.

It is true that, while I spent time merely observing the customs of other men, I found hardly anything about which to be confident and that I noticed there was about as much diversity as I had earlier found among the opinions of philosophers. Thus the greatest profit I derived from this was that on realizing that many things, although they seemed very extravagant and ridiculous to us, did not cease being commonly accepted and approved by other great peoples, I learned to believe nothing very firmly concerning what I had been persuaded to believe only by example and custom; and thus I gradually freed myself from many errors that can darken our natural light and render us less able to listen to reason. But after spending many years thus studying in the book of the world and in trying to gain experience, I made up my mind one day also to study myself and to spend all the powers of my mind in choosing the ways which I ought to follow. For me this procedure was much more successful, it seems, than if I had never left either my country or my books.

11

PART TWO

I was in Germany then, where the wars—which are still continuing there[2]—called me; and while I was returning to the army from the coronation of the emperor, the onset of winter held me up in quarters where, finding no conversation with which to be diverted and, fortunately, otherwise having no worries or passions which troubled me, I remained for a whole day by myself in a small stove-heated room,[3] where I had complete leisure for communing with my thoughts. Among them, one of the first that I thought of considering was that often there is less perfection in works made of several pieces and in works made by the hands of several masters than in those works on which but one master has worked. Thus one sees that buildings undertaken and completed by a single architect are commonly more beautiful and better ordered than those that several architects have tried to patch up, using old walls that had been built for other purposes. Thus these ancient cities that were once merely straggling villages and have become in the course of time great cities are commonly quite poorly laid out, compared to those well-ordered towns that an engineer lays out on a vacant plain as it suits his fancy. And although, upon considering one by one buildings in the former class of towns, one finds as much art or more than one finds in buildings of the latter class of towns, still, upon seeing how the buildings are arranged—here a large one, there a small one—and how they make the streets crooked and uneven, one will say that it is chance more than the will of

12

[2] The Thirty Years War (1618-1648).

[3] There is no need to allege that Descartes sat in or on a stove. A *poêle* is simply a room heated by an earthenware stove. Cf. E. Gilson, *Discours de la méthode: texte et commentaire* (Paris: Vrin, 1976), p. 157.

some men using their reason that has arranged them thus. And if one considers that there have nevertheless always been officials responsible for seeing that private buildings be made to serve as an ornament for the public, one will know that it is difficult to produce a finely executed product by laboring only on the works of others. Thus I imagined that people who, having once been half savages and having been civilized only gradually, have made their laws only to the extent that the inconvenience caused by crimes and quarrels forced them to do so, would not be as well ordered as those who, from the very beginning of their coming together, have followed the fundamental precepts of some prudent legislator. Thus it is quite certain that the state of the true religion, whose ordinances were fixed by God alone, ought to be incomparably better governed than all the others. And, speaking of matters human, I believe that if Sparta flourished greatly in the past, it was not because of the goodness of each of its laws taken by itself, since some of them were very strange and even contrary to good morals, but because, having been invented by only one person, they all tended toward the same goal. And thus I thought that book learning, at least the kind whose arguments are merely probable and have no demonstrations—having been built up from and enlarged gradually by the opinions of many different people—does not draw as near to the truth as the simple reasonings that can be made naturally by a man of good sense concerning what he encounters. 13 And thus again I thought that, given the fact that we were all children before being adults and that for a long time it was our lot to be governed by our appetites and our teachers (both were often in conflict with one another, and perhaps none of them consistently gave us the best advice), it is almost impossible for our judgments to be as pure or solid as they would have been had we the full use of our reason from the moment of our birth and had we never been led by anything but our reason.

It is true that one does not see people pulling down all the houses in a city simply to rebuild them some other way and to make the streets more attractive; but one does see that several people do tear down their own houses in order to rebuild them, and that even in some cases they are forced to do so when their houses are in danger of collapsing and the foundations are not very steadfast. Taking this example to heart, I was persuaded that it was not really likely that a single individual might plan to reform a state by changing everything from the very foundations and by toppling it in order to set it up again; nor even also to reform all of the sciences or the order established in the schools for teaching them; but that I could not do better than to try once and for all to get all the beliefs I had accepted from birth out of my mind, so that once I have reconciled them with reason I might again set up either other, better ones or even the same ones. And I firmly believed that by this means I would 14 succeed in conducting my life much better than were I to build only on old foundations or to lean only on the principles of which I permitted myself to be persuaded in my youth without ever having examined whether or not they were true. For although I noticed various difficulties in this operation, still they did not seem irremediable or comparable to those difficulties arising in the reformation of the least things which affect the public. These great bodies are too difficult to raise up once they have been knocked down or even to maintain once they have been shaken; and their falls can only be very violent. Now as to their imperfections, if they have any (and the mere fact of their diversity suffices to assure one that many of them are imperfect), usage has doubtlessly mitigated them and has even imperceptibly averted or corrected a great number of them, for which deliberate foresight could not have provided so well. And finally, these imperfections are almost always more tolerable than what it takes to change them; simliarly, the great roads that wind around mountains become gradually so level and suitable by dint of being used frequently, that it is much better to follow these roads than to try to go by a more direct route, climbing over rocks and descending to the bottom of precipices.

This is why I could not approve of all of those trouble-making and quarrelsome types who, called neither by birth nor by fortune to manage public affairs, never cease in their imagination to effect some new reformation. And if I thought there were the slightest thing in this essay by means of which one might suspect me of such folly, I would be very sorry to permit its publication. My plan has never been more than to try to reform my own thoughts and to build upon a foundation which is completely my own. And if, my work having sufficiently pleased me, I show it to you here as a model, it is not for that reason that I wish to advise anyone to imitate it. Perhaps those with whom God has better shared his graces have more lofty plans; but I fear that this plan of mine may already be too arduous for many. The single resolution to detach oneself from all the beliefs one has once accepted as true is not an example that everyone ought to follow; and the world consists almost completely of but two kinds of people and for these two kinds it is not at all suitable: namely those who, believing themselves more capable than they really are, cannot help making premature judgments and do not have enough patience to conduct their thoughts in an orderly manner; thus, if they once take the liberty to doubt the principles they have accepted and to keep away from the common path, they could never keep to the path one must take in order to go in a more forward direction—they would remain lost all of their lives. Now as for those people who have enough reason or modesty to judge that they are less capable to distinguish the true from the false than are others by whom they can be instructed, they ought to content themselves more with following the opinions of these others than to look for better opinions on their own.

And for my part, I would unquestionably have been among these latter   16
persons were I to have had only one master or had never seen the differences that have always existed among the opinions of the most learned
people. But having learned since my school days that one cannot imagine
anything so strange or unbelievable that it has not been said by some philosopher, and since then, during my travels, having acknowledged that
those who have feelings quite contrary to our own are not for that reason
barbarians or savages, but that many of them use their reason as much or
more than we do, and having considered how the very same man with his
very own mind, having been brought up from infancy among the French
or the Germans becomes different from what he would be had he always
lived among the Chinese or among cannibals; and how, even to the fash-
ions of our clothing, the same thing that pleased us ten years ago and that
perhaps might again please us ten years from now seem to us extravagant
and ridiculous. Thus it is more custom and example that persuades us than
certain knowledge, and for all that, the majority opinion is not a proof
worth anything for truths that are a bit difficult to discover, since it is
more likely that only one man has found them than a whole people: I
could find no one whose opinions, it seemed to me, ought to be preferred
over the others, and I found myself constrained to try to lead myself on
my own.

But, like a man who walks alone in the shadows, I resolved to go so
slowly and to use so much circumspection in all things that, if I never   17
advanced but slightly, I would at least avoid falling. Moreover, I did not
wish to begin to reject utterly any of these opinions that could have at
some time slipped into my head without having been placed there by my
reason, until I had already spent sufficient time formulating the outline
of the work I was attempting and searching for the true method of arriving at the knowledge of everything my mind was capable of attaining.

In my younger days I studied, among the parts of philosophy, a bit
of logic, and in mathematics, some geometrical analysis and algebra—
three arts or sciences that seemed destined to contribute something to
my plan. But in examining them, I saw that, in the case of logic, its syl-
logisms and the greater part of its other lessons served more to explain
to someone else what one knows, or even, like the art of Lully,[4] to speak
without judgment concerning matters about which one is ignorant, than
to learn them. And although it contains, in effect, very true and good

[1] Ramon Lull (ca. 1236–1315), Catalan philosopher and Franciscan who wrote in
defense of Christianity against the Moors by attempting to demonstrate the articles
of faith by means of logic. Descartes seems to have encountered a Lullist in
Dordrecht who could hold forth on any subject whatever for long periods of time.
This encounter, more than any direct contact with the writings of Lull, seems to
have colored Descartes' understanding of the "art of Lully" Cf. E. Gilson, *Discours
de la méthode: texte et commentaire*, pp. 185–186.

precepts, nevertheless there are so many others, mixed up with them, that are either harmful or superfluous, that it is almost as hard to separate the latter precepts from the former as it is to draw a Diana or a Minerva from a block of marble that is not yet blocked out. Now, as to the analysis of the ancients and the algebra of the moderns, apart from the fact that they apply only to very abstract matters and seem to have no practical utility, the former is always so subject to the consideration of figures that it cannot exercise the understanding without exhausting the imagination; and in the case of algebra, one is so governed by certain laws and symbols that out of it has been made a confused and obscure art that encumbers the mind instead of a science that cultivates it. That is why I believed it necessary to search for another method that, while embracing the advantages of the three, was free from their defects. And since the multiplicity of laws often provides excuses for vices, so that a state is much better when, having but a few laws, its laws are strictly observed; so, in place of the large number of rules of which logic is composed, I believed that the following four rules would be sufficient, provided I made a firm and constant resolution not even once to fail to observe them:

The first was never to accept anything as true that I did not know evidently to be so; that is, carefully to avoid precipitous judgment and prejudice; and to include nothing more in my judgments than what presented itself to my mind with such clarity and distinctness that I would have no occasion to put it in doubt.

The second, to divide each of the difficulties I was examining into as many parts as possible and as is required to solve them best.

The third, to conduct my thoughts in an orderly fashion, commencing with the simplest and easiest to know objects, to rise gradually, as by degrees, to the knowledge of the most composite things, and even supposing an order among those things that do not naturally precede one another.

And last, everywhere to make enumerations so complete and reviews so general that I would be sure of having omitted nothing.

These long chains of reasoning, each of them simple and easy, that geometricians commonly use to attain their most difficult demonstrations, have given me an occasion for imagining that all the things that can fall within human knowledge follow one another in the same way and that, provided only that one abstain from accepting anything as true that is not true, and that one always maintains the order to be followed in deducing the ones from the other, there is nothing so far distant that one cannot finally reach nor so hidden that one cannot discover. And I was not very worried about seeking which of them it would be necessary to

begin with; for I already knew that it was with the simplest and easiest
to know. And considering that all of those who have already searched
for truth in the sciences, only the mathematicians were able to find demonstrations, that is, certain and evident reasons, I did not doubt that it
was with these same starting points that they had conducted their exam-
inations; although I expected no further usefulness from them, except
that they would accustom my mind to feed upon truths and not to be
content with false reasons. But in all of this it was not my plan to try
to learn all of the specific sciences that commonly are called mathematics; and seeing that, even though their objects differed, they did not   20
cease to be in accord with one another, in that they consider only the
various relations or proportions which are in these objects, I believed it
would be more worthwhile were I to examine only these proportions in
a general way, and to suppose them to be in concrete objects only to
the extent that these objects aid me in making it easier to acquire knowledge of these proportions, and also without in any way binding these
proportions to those objects, so that later one can apply them all the
better to everything else to which they might pertain. Now, having noticed that, in order to know these proportions, I occasionally needed to
consider each of them individually, and sometimes only to remember
them, or to gather up several of them together, I believed that, to consider them better in particular, I ought to suppose them as relations between lines, since I found nothing more simple, nothing that I could
more distinctly represent to my imagination and my senses; but to remember them or to grasp them all together, I would have had to explicate them by means of certain symbols, the shortest ones possible; and
by this means I would borrow all of the better aspects of geometrical
analysis and algebra, and I would correct all the defects of the one by
means of the other.

In effect, I dare say that the exact observance of these few precepts
that I have chosen gave me such a facility for disentangling all the problems to which these two sciences tended, that in the two or three months
I spent examining them, having begun from the simplest and most general—and each truth that I found being a rule which I later used to find   21
others—I not only succeeded in several problem areas that I had once
judged very difficult, but it also seemed to me toward the end that I
could determine, even in those problem areas where I was ignorant, by
what means and how far it would be possible to resolve them. In this
perhaps I shall not seem to you to be too vain, if you were to consider
that, there being but one truth for each thing, anyone who finds it knows
as much as one can know about that thing; and that, for example, a child
given lessons in arithmetic, having made one addition in accordance with
its rules, can be assured of having found everything the human mind can
find bearing on the sum he has examined. For thus, the method, which

teaches one to follow the true order and to enumerate exactly all the circumstances of what one is seeking, contains everything that gives certainty to the rules of arithmetic.

But what pleased me the most about this method was that by means of
it I was assured of using my reason in everything, if not perfectly, then
at least as best as I can. Moreover, I felt that in practicing this method my
mind was gradually getting into the habit of conceiving its object more
rigorously and more distinctly and that, not having subjected it to any
particular matter, I promised myself to apply the method just as profitably to the problems of the other sciences, as I had done to problems in
algebra. Not that, on account of this, I dared immediately to undertake
an examination of whatever presented itself; for even that would have
been contrary to the order prescribed by the method. But having noticed
22   that their principles must all be borrowed from philosophy, in which I
still found nothing certain, I thought that I ought, above all, to try to establish therein something certain; and I thought that, this being the
most important thing in the world, where precipitous judgment and
prejudice were most to be feared, I ought not to have tried to succeed
at doing so until I had reached a much more mature age than merely
twenty-three, which I was then; and I thought that I should previously
spend much time preparing myself, as much in rooting out of my mind
all the wrong opinions that I had accepted before that time as in accumu-
lating many experiences—later to be the stuff of my reasonings—and in
always exercising myself in the method I had prescribed for myself so
as to be stronger and stronger in its use.

Part Three

Now just as it is not enough, before beginning to rebuild the house
where one lives, to pull it down, to make provisions for materials and
architects, or to take a try at architecture for oneself, and also to have
carefully worked out the floorplan; one must provide for something else
in addition, namely where one can be conveniently sheltered while working on the other building; so too, in order not to remain irresolute in my
actions while reason requires me to be so in my judgments, and in order
not to cease living during that time as happily as possible, I formulated a
provisional code of morals, which consisted of but three or four maxims,
that I want to share with you.
23   The first was to obey the laws and the customs of my country, firmly
holding on to the religion in which, by God's grace, I was instructed
from childhood, and governing myself in all other things according to
the most moderate opinions and those furthest from excess that were

commonly accepted in practice by the most sensible of those people with whom I would have to live. For, already beginning to count my own opinions as nothing, since I wished to remit all opinions to examination, I was assured that I could not do better than to follow the opinions of those who were the most sensible. And although there may perhaps be people among the Persians and the Chinese just as sensible as there are among ourselves, it seemed to me that the most useful course of action was to rule myself in accordance with those with whom I had to live, and that, to know their true opinions, I ought to observe what they do rather than what they say, not only because in the corruption of our morals there are few people who are willing to say all they believe, but also because many do not know what they believe; for, given that the action of thought by which one believes something is different from that by which one knows that one believes it, the one often occurs without the other. And among several opinions held equally, I would choose only the most moderate, not only because it is always the most suitable for action and probably the best (every excess usually being bad), but also so as to stray less, in case I am mistaken, from the true road—having chosen one of the two extremes—when it was the other one I should have followed. And in particular I placed among the excesses all of the promises by which one curtails something of one's freedom. Not that I   24
disapprove of laws that, to remedy the inconstancy of weak minds, permit (when one has a good plan or even, for the security of commerce, a plan that is only indifferent) one to make vows or contracts that oblige one to persevere in them; but because I have seen nothing in the world that remains always in the same state and because, for my part, I promised to perfect my judgments more and more, and not to render them worse, I would have believed I committed a grave indiscretion against good sense if, having once approved of something, I obligated myself to take it to be good at a later time when perhaps it would have ceased to be so or when I would have ceased judging it to be good.

My second maxim was to be as firm and resolute in my actions as I could be, and to follow with no less constancy the most doubtful opinions, once I have decided on them, than if they were very certain. In this I would imitate travelers who, finding themselves lost in a forest, ought not wander this way and that, or, what is worse, remain in one place, but ought always walk as straight a line as they can in one direction and not change course for feeble reasons, even if at the outset it was perhaps only chance that made them choose it; for by this means, if they are not going where they wish, they will finally arrive at least somewhere where they   25
probably will be better off than in the middle of a forest. And thus the actions of life often tolerating no delay, it is a very certain truth that, when it is not in our power to discern the truest opinions, we ought to

follow the most probable; and even if we observe no more probability in some than in others, nevertheless we ought to fix ourselves on some of them and later consider them no longer as doubtful, insofar as they relate to practical affairs, but as very true and very certain, since reason, which has caused us to make this determination, is itself of the same sort. And this insight was capable, from that point onward, of freeing me from the repentence and remorse that commonly agitate the consciences of these frail and irresolute minds that allow themselves to go about with inconstancy, treating things as if they were good, only to judge them later to be bad.

My third maxim was always to try to conquer myself rather than fortune, to change my desires rather than the order of the world; and generally to become accustomed to believing that there is nothing that is utterly within our power, except for our thoughts, so that, after having done our best regarding things external to us, everything that fails to bring us success, from our point of view, is absolutely impossible. And this principle alone seemed sufficient to stop me from desiring anything in the future that I would not acquire, and thus seemed sufficient to make me contented. For, our will tending naturally to desire only what our intellect represents to it as in some way possible, it is certain that, if we consider all of the goods that are outside us as equally beyond our power, we should have no more regrets about lacking what seems owed to us at birth, when we are deprived of them through no fault of our own, than we should have for not possessing the kingdoms of China or Mexico. Thus, making a virtue of necessity, as they say, we shall no more desire to be healthy if we are sick, or to be free if we are in prison, than we would desire to have a body made of matter as incorruptible as diamonds, or wings with which to fly like birds. But I confess that long exercise is needed as well as frequently repeated meditation in order to become accustomed to looking at everything from this point of view; and I believe that in this principally lay the secret of the philosophers who at one time were able to free themselves from fortune's domination and who could, despite their sorrows and their poverty, rival their gods in their happiness. For occupying themselves ceaselessly with considering the limits prescribed to them by nature, they so perfectly persuaded themselves that nothing was in their power save their own thoughts, that this alone was sufficient to stop them from having feelings about any other objects; and they controlled their thoughts so absolutely, that they thereby had some reason for judging themselves richer, more powerful, freer, and happier than those other men who, not having this philosophy—however favored by nature and fortune as they may be—never controlled everything they wished to control.

Finally, to conclude this code of morals, I thought it advisable to review the various occupations that men take up in this life, so as to try to choose the best one; and, not wanting to say anything about the occupations of others, I believed I could not do better than to continue in the occupation I was in at that time, namely cultivating my reason all my life and advancing, as best as I could, in the knowledge of truth, following the method I had prescribed to myself. I had met with such intense satisfaction, since the time I had begun to make use of this method, that I did not believe one could receive sweeter or more innocent satisfaction in this life; and, discovering every day by its means some truths that seemed important to me and commonly ignored by other men, I had a satisfaction that so filled my mind that nothing else was of any consequence to me. In addition, the three preceding maxims were founded merely on the plan I had of continuing my self-instruction; for since God has given each of us a certain light by which to distinguish the true from the false, I should not believe I ought to be content for a single moment with the opinions of others, had I not proposed to use my own judgment to examine them while there was time; and I should not have been able to be free of scruple in following these opinions, had I not hoped I would not waste the opportunity thereby of finding better ones, in case there were better ones. And finally, I could not have curbed my desires or have been contented, had I not followed a road by which, believing I was assured of acquiring all the knowledge of which I was capable, I believed I was assured of acquiring by the same means all the true goods that would ever be in my power; given that our will neither pursues nor flees an object unless our intellect represents that object to the will as either good or bad, it suffices to judge well in order to do well, and to judge as best one can, in order also to do one's best, that is, to acquire all the virtues and, along with them, all the other goods that one can acquire; and while one is certain that this is the case, one could not fail to be contented.

After having assured myself of these maxims and having put them aside, along with the truths of the faith, which have always held first place in my set of beliefs, I judged that, as far as the rest of my opinions were concerned, I could freely undertake to rid myself of them. And insofar as I hoped I could be more successful in social interchange than in remaining any longer shut up in the small stove-heated room where I had had all of these thoughts, I set out again on my travels, the winter not yet completely over. And in all the following nine years I did nothing but wander here and there about the world, trying to be more a spectator than an actor in all the comedies that were being played out there; and reflecting particularly in each matter on what might render it suspect and give us occasion for error, I meanwhile rooted out from my mind all the errors that had been able to creep in undetected. Not that I was thereby aping the sceptics who doubt merely for the sake of doubting and put on

28

29

the affectation of perpetual indecision; for, on the contrary, my entire plan tended simply to give me assurance and to reject shifting ground and sand so as to find rock or clay. In this I was quite successful, it seems to me, inasmuch as, trying to discover the falsity or uncertainty of the propositions I was examining—not by feeble conjectures but by clear and certain reasonings—I never found anything that was so doubtful that I could not draw some rather certain conclusion from it, even if it were merely that it contained nothing certain. And just as in tearing down an old house, one usually saves the wreckage for use in building the new house, similarly, in destroying all of those opinions that I judged to be poorly supported, I made various observations and acquired many experiences that I later found useful in establishing opinions that were more certain. Moreover, I continued to practice the method I had prescribed for myself; for besides taking care generally to conduct all my thoughts according to the rules of this method, from time to time I set aside a few hours that I spent in practicing the method on mathematical problems, or even in various other problems that I could render similar to those of mathematics, by detaching them from all the principles of the other sciences, which I did not find to be sufficiently firm, as you will observe I have done in several cases that are explained in this volume. And thus, 30 without living any other way in outward appearance than those who, having no other task but living sweet and innocent lives, are eager to separate pleasures from vices and who, to enjoy their leisure without becoming bored, engage in all sorts of honorable diversions, I did not cease to pursue my plan and to profit in the knowledge of the truth, perhaps more than if I had done nothing but read books or keep company with men of letters.

All the same, these nine years slipped away before I had as yet taken any stand regarding the difficulties commonly debated by learned men, or had I begun to seek the foundations of any philosophy that was more certain than the commonly accepted one. And the example of many excellent minds, which, having already had this plan, appeared to me not to have succeeded, made me conjure thoughts of so many difficulties that perhaps I should not yet have dared to try it if I had not seen that some people had already passed the rumor around that I had already succeeded. I cannot say on what they based this opinion; and if I have contributed anything to this by my conversations, it must have been more because I admitted what I did not know more ingenuously than do those who have studied only a little, and perhaps also because I showed the reasons I had for doubting many of the things that other people regard as certain, than because I was boasting of any knowledge. But being decent enough not to want someone to take me for something other than I was, I thought it necessary to try by every means to make myself worthy

of the reputation bestowed upon me; and it is exactly eight years since   31
this desire made me resolve to take my leave of all those places where I
could have acquaintances, and to retire here, in a country where the long
duration of the war has established such well-ordered discipline that the
armies quartered there seem to be there solely for the purpose of guaran-
teeing the enjoyment of the fruits of peace with even greater security, and
where among the crowds of a great and very busy people and more concerned with their own affairs than curious about the affairs of others, I
have been able to live as solitary and as retired a life as I could in the re-
motest deserts—but without lacking any of the amenities that are to be
found in the most populous cities.

Part Four

I do not know whether I ought to tell you about the first meditations
I made there; for they are so metaphysical and so out of the ordinary,
that perhaps they would not be to everyone's liking. Nevertheless, so
that one might be able to judge whether the foundations I have laid are
sufficiently firm, I am in some sense forced to speak. For a long time I
have noticed that in moral matters one must sometimes follow opinions
that one knows are quite uncertain, just as if they were indubitable, as
has been said above; but since then I desired to attend only to the search
for truth, I thought it necessary that I do exactly the opposite, and that I
reject as absolutely false everything in which I could imagine the least
doubt, so as to see whether, after this process, anything in my set of beliefs remains that is entirely indubitable. Thus, since our senses sometimes   32
deceive us, I decided to suppose that nothing was exactly as our senses
would have us imagine. And since there are men who err in reasoning,
even in the simplest matters in geometry, and commit paralogisms, judging that I was just as prone to err as the next man, I rejected as false all
the reasonings that I had previously taken for demonstrations. And finally,
taking into account the fact that the same thoughts we have when we
are awake can also come to us when we are asleep, without any of the
latter thoughts being true, I resolved to pretend that everything that
had ever entered my mind was no more true than the illusions of my
dreams. But immediately afterward I noticed that, during the time I
wanted thus to think that everything was false, it was necessary that I,
who thought thus, be something. And noticing that this truth—*I think,
therefore I am*—was so firm and so certain that the most extravagant
suppositions of the sceptics were unable to shake it, I judged that I
could accept it without scruple as the first principle of the philosophy
I was seeking.

Then, examining with attention what I was, and seeing that I could pretend that I had no body and that there was no world nor any place where I was, but that I could not pretend, on that account, that I did not exist; and that, on the contrary, from the very fact that I thought about doubting the truth of other things, it followed very evidently and very

33   certainly that I existed. On the other hand, had I simply stopped thinking, even if all the rest of what I have ever imagined were true, I would have no reason to believe that I existed. From this I knew that I was a substance the whole essence or nature of which was merely to think, and which, in order to exist, needed no place and depended on no material thing. Thus this "I," that is, the soul through which I am what I am, is entirely distinct from the body, and is even easier to know than the body, and even if there were no body, the soul would not cease to be all that it is.

After this, I considered in a general way what is needed for a proposition to be true and certain; for since I had just found a proposition that I knew was true, I thought I ought also know in what this certitude consists. And having noticed that there is nothing in all of this—*I think, therefore I am*—that assures me that I am uttering the truth, except that I see very clearly that, in order to think, one must exist; I judged that I could take as a general rule that the things we conceive very clearly and very distinctly are all true, but that there only remains some difficulty in properly discerning which are the ones that we distinctly conceive.

Following this, reflecting upon the fact that I doubted and that, as a consequence, my being was not utterly perfect (for I saw clearly that it is a greater perfection to know than to doubt), I decided to search for the source from which I had learned to think of a thing more perfect

34   than myself; and I readily knew that this ought to originate from some nature that was in effect more perfect. As to those thoughts of mine that were of many other things outside me—such as the sky, the earth, light, heat, and a thousand other things—I was not quite so anxious to know where they came from, since, having noticed nothing in them that seemed to me to make them superior to me, I could believe that, if they were true, they were dependencies of my nature, to the extent that it had any perfection; and that if they were not true, I received them from nothing, that is, they were in me because I had some defect. But the same could not hold for the idea of a being more perfect than my own; for the receiving of this idea from nothing is a manifest impossibility; and since it is no less a contradiction that something more perfect should follow from and depend upon something less perfect than that something can come from nothing, I certainly could not obtain it from myself. It thus remained that this idea was placed in me by a nature truly more perfect than I was, and even that it had within itself all the perfections of which I could have any idea, that is, to put my case in a single word, that this

nature was God. To this I added that, since I knew of some perfections that I did not possess, I was not the only being in existence (here, if you please, I shall use freely the language of the School), but that of necessity it must be the case that there is something else more perfect, upon which I depended, and from which I acquired all that I had. For, had I been alone and independent of everything else, so as to have derived from my-   35
self all of that small allotment of perfection I had through participation in the perfect being, I would have been able for the same reason to give myself the remainder of what I knew was lacking in me; and thus I would be infinite, eternal, unchanging, all-knowing, all-powerful—in short, I would have all the perfections I could discern in God. For, following from the reasonings I have just given, to know the nature of God, as far as my own nature was able, I had only to consider each thing about which I found an idea in myself, whether or not it was a perfection to have them, and I was certain that none of those that were marked by any imperfection were in this nature, but that all other perfections were. So I observed that doubt, inconstancy, sadness and the like could not be in him, given the fact that I would have been happy to be exempt from them. Now, over and above that, I had ideas of several sensible and corporeal things; for even supposing that I was dreaming and that everything I saw or imagined was false, I still could not deny that the ideas were not truly in my thought. But since I had already recognized very clearly in my case that intelligent nature is distinct from corporeal nature, taking into consideration that all composition attests to dependence and that dependence is manifestly a defect, I therefore judged that being composed of these two natures cannot be a perfection in God and that, as a consequence, God is not thus composed. But, if there are bodies in the world, or intelligences, or other natures that were not entirely perfect, their   36
being ought to depend on God's power, inasmuch as they cannot subsist without God for a single moment.

After this, I wanted to search for other truths, and, having set before myself the object dealt with by geometricians, which I conceived to be like a continuous body or a space indefinitely extended in length, breadth, and height or depth, divisible into various parts which could have various shapes and sizes and be moved or transposed in all sorts of ways (for all this the geometricians take for granted in their object), I ran through some of their simplest proofs. And having noticed that this great certitude that everyone attributes to them is founded only on the fact that one conceives them evidently conforming to the rule that I mentioned earlier, I also noted that there had been nothing in them that assured me of the existence of their object. For I saw very well that by supposing, for example, a triangle, it is necessary for its three angles to be equal to two right angles; but I did not see anything in all this which would assure me that any triangle existed. On the other hand, returning to an examination of the idea I had of a perfect being, I found that existence was contained in it, in the same way as the fact that its three angles are equal to two right angles is contained in the idea of a triangle, or that, in the case of a sphere, all its parts are equidistant from its center, or even more evidently so; and consequently, it is, at the very least, just as certain that God, who is a perfect being, is or exists, as any demonstration in geometry could be.

37   But what makes many people become persuaded that it is difficult to know this (i.e., the existence of the perfect being), and also even to know what kind of thing their soul is, is that they never lift their minds above sensible things and that they are so much in the habit of thinking about only what they can imagine (which is a particular way of thinking appropriate only for material things), that whatever is not imaginable seems to them to be unintelligible. This is obvious enough from what even the philosophers in the Schools take as a maxim: that there is nothing in the understanding that has not first been in the senses (where obviously the ideas of God and the soul have never been). And it seems to me that those who want to use their imagination to comprehend these things are doing the same as if, to hear sounds or to smell odors, they wanted to use their eyes, except for this difference: the sense of sight assures us no less of the truth of its objects than do the senses of smell or hearing, whereas neither our imagination nor our sense could ever assure us of anything if our understanding did not intervene.

Finally, if there are men who have not yet been sufficiently persuaded of the existence of God and their soul by means of the reasons I have brought forward, I would very much like them to know that all the other things they thought perhaps to be more certain—such as having a body, there being stars and an earth, and the like—are less certain. For although one might have a moral certainty about these things, which is such that it seems outrageous for anyone to doubt it, yet, while it is a question of metaphysical certitude, it seems unreasonable for anyone to deny that there is a sufficient basis for one's not being completely certain about the subject, given that one can, in the same fashion, imagine that while asleep one has a different body and that one sees different stars and a different earth, without any of it being the case. For how does one know that the thoughts that come to us in our dreams are more false than the others, given that often they are no less vivid or express? Let the best minds study this as much as they please, I do not believe they can give any reason that would suffice to remove this doubt, were they not to presuppose the existence of God. For first of all, even what I have already taken for a rule—namely that all the things we very clearly and very distinctly conceive are true—is certain only because God is or exists, and is a perfect

being, and because all that is in us comes from him. Thus it follows that
our ideas or our notions, being real things and coming from God, insofar
as they are clear and distinct, cannot to this extent fail to be true. Thus,
if we have ideas sufficiently often that contain falsity, this can only be
the case with respect to things that have something confused or obscure
about them, since in this regard they participate in nothing; that is, they
are thus in us in such a confusion only because we are not perfect. And
it is evident that there is no less a contradiction that falsity or imperfection, as such, proceed from God, than that truth or perfection proceed   39
from nothing. But if we did not know that all that is real and true in us
comes from a perfect and infinite being, however clear and distinct our
ideas may be, we would have no reason that assured us that they had the
perfection of being true.

But after the knowledge of God and the soul has thus rendered us certain of this rule, it is very easy to know that the dreams we imagine while
asleep ought in no way make us doubt the truth of the thoughts we have
while awake. For if it should happen, even while one is asleep that someone has a very distinct idea, as, for example, when a geometrician invents
a new demonstration, his being asleep does not impede its being true. And
as to the most common error of our dreams, which consists in the fact
that they represent to us various objects in the same way as our exterior
senses do, it is of no importance that it gives us the occasion to question
the truth of such ideas, since they can also deceive us just as often without our being asleep—as when those with jaundice see everything as yel-
low-colored, or when the stars or other distant bodies appear to us a great
deal smaller than they are. In short, whether awake or asleep, we should
never allow ourselves to be persuaded except by the evidence of our reason. And it is to be noted that I said this of our reason, and not of our
imagination or our senses. For, although we see the sun very clearly, we   40
should not on that account judge that it is only as large as we see it; and
we can very well imagine distinctly the head of a lion grafted on the body
of a goat, without necessarily concluding for that reason that there existed a chimera; for reason does not suggest to us that what we thus see
or imagine is true. But it does suggest to us that all our ideas or notions
ought to have some foundation in truth; for it would not be possible that
God, who is all perfect and entirely truthful, would have put them in us
without that. And because our reasonings are never so evident nor so
complete while we are asleep as they are while we are awake, even though
our imaginations are sometimes just as, or even more, vivid and express
when asleep, reason also suggests to us that our thoughts are unable all
to be true, since we are not all-perfect; what truth there is in them ought
infallibly to be found in those we have when awake rather than those we
have in our dreams.

PART FIVE

I would be quite happy to continue and to show here the whole chain of other truths that I had deduced from these first ones. But since, to that end, it would now be necessary for me to speak about many problems that are a matter of controversy among the learned, with whom I do not want to get into a scuffle, I believe it would be better for me to abstain and to state only in a general way what they are, so that it might be left to the most wise to judge whether it be useful for the public to be more
41 informed as to the particulars. I have always remained firm in my resolve not to suppose any principle but the one I just used to demonstrate the existence of God and the soul, and to take nothing to be true that does not seem to me clearer and more certain than have the demonstrations of the geometricians been previously. And still I dare say not only that I have found the means of satisfying myself in a short time regarding all the main difficulties commonly treated in philosophy, but also that I have noted certain laws that God has so established in nature and has impressed in our souls such notions of these laws that, after having reflected sufficiently, we cannot deny that they are strictly adhered to in everything that exists or occurs in the world. Now, in considering the chain of these laws, it seems that I have discovered several truths more useful and more important than all I had previously learned or even hoped to learn.

But because I have tried to explain these principles in a treatise that certain considerations kept me from publishing,[5] I could not make them better known than by declaring here in summary form what the treatise contains. I had intended to include in it everything I thought I knew, before writing it down, concerning the nature of material things. But just as painters who, unable to represent equally on a flat surface all the various sides of a solid body, choose one of the principal sides which they
42 place alone in the light of day, and, darkening with shadows all the rest, make them appear only as they can be seen while someone is looking at the principal side; just so, fearing I could not put in my discourse all that I had thought on the matter, I tried simply to speak at length about what I conceived with respect to light; then, at the proper time, to add something about the sun and the fixed stars, since light originates almost entirely from them; something about the heavens since they transmit light; about the planets, comets, and the earth since they reflect light; and particularly, about all terrestrial bodies, since they are either colored,

---

[5] Descartes' *Le Monde* (*The World*). See René Descartes, *Le Monde ou Traité de la lumière*, translation and introduction by Michael Sean Mahoney. (New York: Abaris Books, Inc., 1979). One of the considerations preventing the publication of *Le Monde* was the trial in 1633 of Galileo by the Holy Office in Rome.

transparent, or luminous; and finally, about man, since he is the observer of all this. But, to put all these things in a slightly softer light and to be able to say more freely what I have judged in these matters, without being obliged either to follow or to refute the opinions accepted among the learned, I here resolved to leave all this world to their disputes and to speak only of what would happen in a new world, were God now to create enough matter to make it up, somewhere in imaginary space, and if he were to put in motion variously and without order the different parts of this matter, so that he concocted as confused a chaos as the poets could ever imagine and that later he did no more than apply his ordinary conserving activity to nature, letting nature act in accordance with the laws he has established. Thus, first, I described this matter and tried to represent it such that there is nothing in the world, it seems to me, more clear and more intelligible, with the exception of what has already been said about God and the soul; for I even supposed explicitly that there was in 　　**43** it none of those forms or qualities about which disputes occur in the Schools, nor generally anything the knowledge of which was not so natural to our souls that one cannot even pretend to ignore it. Moreover, I showed what were the laws of nature; and without supporting my reasons on any other principle but the infinite perfections of God, I tried to demonstrate all the laws about which one might have been able to doubt and to show that they were such that, even if God had created several worlds, there could have been none in which these laws failed to be observed. Next, I showed how most of the matter of this chaos must, in observance of these laws, dispose and arrange itself in a certain way that makes it similar to our heavens; how, meanwhile, some of its parts should form an earth; others, planets and comets; and still others a sun and fixed stars. And here, dwelling on the subject of light, I explained at great length what this light was that ought to be found in the sun and the stars, and how thence it coursed in an instant the immense stretches of celestial space, and how it was reflected from the planets and comets to the earth. To that I added also several things touching on the substance, position, movements, and all the various qualities of these heavens and stars; and as a result I thought that I said enough on the matter to show that there is nothing to be mentioned in this world which should not, or at least could not, appear utterly similar to the world I have just described. 　　**44** Next I went on to speak particularly of the earth: how, although I had expressly supposed that God had not put any weight in the matter out of which the earth was composed, still none of its parts would cease to tend precisely toward its center; how, having water and air on its surface, the disposition of the heavens, the stars, and principally the moon ought to cause an ebb and flow that would be similar in all respects to what we observe in our own seas, and, what is more, a certain coursing—as much of the water as of the air—from east to west, such as is observed between the tropics; how mountains, seas, springs, and rivers could be formed naturally therein and how metals could have made their way naturally into mines; how plants could have grown naturally in the fields and generally how all the bodies one calls mixed or composed could have been engendered naturally. And, among other things, since over and above the stars I know of nothing else in the world that produces light except fire, I tried to make clearly understood all that belonged to its nature, how it occurs, how it feeds, how sometimes there is only heat but no light, and sometimes only light but no heat; how it can introduce various colors into various bodies, and the same for various other qualties; how it melts some things and hardens others; how it can consume nearly everything or turn it into ashes and smoke; and finally, how from these ashes, merely by the violence of its action, it produces glass; for since this transmutation of ashes into glass seems to me to be as awesome as any other event in nature, I took particular pleasure in describing it.

45

Yet I did not want to infer from all these things that this world has been created in the manner I described, for it is much more likely that, from the very beginning, God made it such as it was supposed to be. But it is certain (and this is an opinion commonly held among theologians) that the action by which God conserves the world is precisely the same action by which he created it; so that even if he had never given it, at the beginning, any other form but that of chaos, provided he established the laws of nature and applied his conserving activity to make nature function just as it does ordinarily, one can believe—without belittling the miracle of creation—that by such activity alone all the things that are purely material could have been able, as time went on, to make themselves just as we now see them. And their nature is much easier to conceive, when one sees them gradually coming to be in this manner, than when one considers them only in their completed state.

From the description of inanimate bodies and plants I passed to the description of animals and particularly of human beings. But since I had insufficient knowledge in this area to speak in the same way as I did in the rest, that is in demonstrating effects by their causes and showing from what seeds and in what way nature should produce them, I contented myself with supposing that God formed the body of a man entirely similar to one of ours, as much as in the outward shape of its members as in the internal arrangement of its organs, without making it out of any material other than the type I had described, and without putting in it, at the start, any rational soul, or anything else to serve as a vegetative or sensitive soul, but merely exciting in the man's heart one of those fires without light that I had already explained, and which, having no other

46

nature but that which heats up hay when it has been bundled up before drying, or which boils new wines while they are left to ferment on the stalk. For examining the functions that could, consequently, be in this body, I found precisely all the things that could be in us without our thinking about them, and hence without our soul (that is, that part distinct from the body of which I have said before that its nature is only to think) contributing anything to them, and these are all the same so that one can say that nonrational animals resemble us. But I could not on that account find any of those things that, being dependent on thought, are the only things that belong to us insofar as we are men, although I found them all later when I had supposed that God created a rational soul and that he joined it to this body in the particular way I have described.

But so that one could see the way in which I treated this matter, I want to place here the explanation of the movement of the heart and arteries; because this is the first and most general movement that one observes in animals, one easily judges on the basis of it what one ought to think regarding all the rest. So that one might have less difficulty in understanding what I shall say on the matter, I would like those who are not versed in anatomy to take the trouble, before reading this, to have dissected in their presence the heart of a large animal that has lungs (for it is in many respects sufficiently similar to that of a man), and to be shown the two chambers or ventricles in the heart. First, the one on the right side of the heart, into which two very large tubes lead, namely the *vena cava*, which is the main receptable for blood, somewhat like the trunk of a tree all of whose other veins are branches, and the *vena arteriosa* (which has been rather ill-named, since it is, after all, an artery), which, taking its origin from the heart, breaks up, on leaving the heart, into several branches that are spread all throughout the lungs. Now the chamber on the left side, into which two tubes lead in the very same fashion, which are just as large or larger than the other tubes: namely, the *arteria venosa* (which also has been ill-named, since it is only a vein), which comes from the lungs where it is divided into many branches interlaced with those of the *vena arteriosa* and with those in the passageway called the windpipe, through which enters the air one breathes; and the aorta which, on leaving the heart, sends its branches all over the body. I would also like to have one be carefully shown the eleven little valves that, like so many little doors, open and shut the four openings in the two ventricles: namely, three at the entrance to the *vena cava*, where they are so disposed that they cannot stop the blood it contains from flowing into the right ventricle of the heart, and yet stop any of it from being able to leave the ventricle; three at the entrance to the *vena arteriosa* which, being disposed in just the opposite way, readily allow

47

48

the blood in this ventricle to pass into the lungs, but do not allow any blood in the lungs to return to this ventricle; two others at the entrance to the *arteria venosa*, which let blood flow from the lungs to the left ventricle of the heart but resist its returning; and three at the entrance to the aorta that permit blood to leave the heart but stop it from returning. And there is no need to search for any other reason for the number of valves except that the opening of the *arteria venosa*, being oval-shaped because of its location, can be conveniently closed with two, while the others, being round, can better be closed with three. Moreover, I would like people to consider that the aorta and the *vena arteriosa* are of a much sturdier and firmer constitution that the *arteria venosa* and the *vena cava;* and that these last two are enlarged before entering the heart and form, as it were, sacks, called the auricles of the heart, which are made of flesh similar to that of the heart; and that there is always more heat in the heart than anywhere else in the body; and, finally, that this heat is able to bring it about that, if a drop of blood should enter its ventricles, the blood ex-

49   pands forthwith and is dilated, just as all liquids generally do, when one lets them fall drop by drop in a very hot vessel.

For, after that, I have no need of saying anything else to explain the movement of the heart, except that, while its ventricles are not full of blood, blood necessarily flows from the *vena cava* into the right ventricle and from the *arteria venosa* into the left ventricle—given that these two vessels are always full, and their apertures, which open toward the heart, cannot then be shut. But as soon as two drops of blood have thus entered the heart, one into each ventricle, these drops—which can only be very large, because the openings through which they enter are very large and the vessels whence they come are quite full of blood—are rarefied and dilated, because of the heat they find there, by means of which, making the whole heart inflate, they push and close the five little doors that are at the entrances of the two vessels whence they come, impeding any further flow of blood into the heart; and continuing to be more and more rarified, they push and open the six other little doors that are at the entrances of the two other vessels by which they leave; by this means they inflate all the branches of the *vena arteriosa* and the aorta, almost at the same instant as the heart, which immediately afterward contracts—as do the arteries too—since the blood that has entered there gets cooled again and their six small doors close, and the five little doors of the *vena cava*

50   and the *arteria venosa* are opened up again and grant passage to two other drops of blood which immediately inflate again the heart and the arteries, the same as before. And because the blood that thus enters the heart passes by the two sacks called auricles, so it follows that their movement is contrary to the heart's: they are deflated while the heart is inflated. For the rest (so that those who do not know the force of mathematical dem-

onstrations and are not in the habit of distinguishing true reasons from apparent reasons should not venture to deny this without examining it), I want to put them on notice that this movement that I have been explaining follows just as necessarily from the mere disposition of the organs that can be seen in the heart by the unaided eye and from the heat that can be felt with the fingers, and from the nature of blood which can be known through experience, as do the motions of a clock from the force, placement, and shape of its counterweights and its wheels.

But if one asks how it is that blood in the veins is not exhausted, in flowing continually into the heart, and how it is that the arteries are not overly full, since all the blood that passes through the heart goes there, I need only answer with what has already been written by an English physician,[6] to whom must be given homage for having broken the ice in this area, and who was the first to have taught that there are many small passages at the extremities of the arteries through which the blood they receive enters in the small branches of the veins, from which it heads once more for the heart, so that its course is merely a perpetual circulation. He proves this very effectively from the common experience of surgeons, who, on binding an arm moderately tightly above the spot where they opened the vein, cause blood to flow out in even greater abundance than if they had not bound the arm. And the opposite would happen if they bound the arm below, between the hand and the opening, or if they tied it above very tightly, because it is clear that a tourniquet tied moderately, being able to stop blood already in the arm from returning to the heart through the veins, does not on that account stop any new blood that is always coming in from the arteries, since they are located below the veins, and their valves, being tougher, are less easy to press, and since the blood coming from the heart tends to pass with greater force through the arteries toward the hand than it does on returning to the heart through the veins. Because the blood flows from the arm through the opening in one of the veins, there must necessarily be some passages below the tourniquet, that is, toward the extremities of the arm, through which it can come from the arteries. He also proves quite effectively what he says regarding the flow of blood, first, through certain small valves that are so disposed in various places along the length of the veins that they do not allow blood to pass from the middle of the body to the extremities, but only to return from the extremities toward the heart; and, second, through experience which shows that all the blood in the

51

---

[6] William Harvey (1578–1657), English physiologist who demonstrated the function of the heart as a kind of pump and the complete circulation of blood throughout the body. His most important work is *Anatomical Exercises on the Motion of the Heart and Blood* (1628)

body can flow out of it in a very short amount of time through just one artery if it is cut open, even if it is very tightly bound quite close to the heart, and cut open between the heart and the tourniquet, so that one could not have any basis for imagining that the blood left from anywhere but the heart.

But there are many other things that attest to the fact that the true cause of this movement of blood is what I said it is. First, the difference that one notices between blood leaving the veins and blood leaving the arteries can arise only from the fact that it is rarified and, so to speak, distilled, in passing through the heart; it is more subtle, more lively, and warmer just after having come out of the heart, that is, while it is in the arteries, than it is shortly before it enters the heart, that is, while it is in the veins. If one takes a good look, one will find that this difference appears more clearly near the heart and not so much in those places furthest removed from the heart. Now the toughness of these valves, which compose the *vena arteriosa* and the aorta, shows quite well that the blood bangs against them with more force than it does against the veins. And why are the left ventricle of the heart and the aorta more spacious and larger than the right ventricle and the *vena arteriosa*, unless it is that the blood in the *arteria venosa*, having been only in the lungs after having passed through the heart, is more subtle and is more forcefully and easily rarified than what comes immediately from the *vena cava*? And what can physicians divine from taking the pulse, if they do not know that, as the blood changes its nature, it can be rarified by the heat of the heart with more or less strength, with more or less liveliness than before? And if one examines how this heat is communicated to the other members, must one not admit that it is by means of the blood which, on passing through the heart, is warmed and from that point is spread throughout the whole body? It follows from this that if one removes the blood from some part, one also removes the heat; and even if the heart were as intensely hot as a piece of glowing iron, it would not be enough to heat the feet and hands as it does, unless it continuously sent new supplies of blood. Now one also knows from this that the true purpose of respiration is to bring sufficient quantities of fresh air into the lungs, to cause the blood, which leaves the right ventricle of the heart where it was rarified and, as it were, changed into vapors, to be condensed and to be converted once again into blood, before returning to the left ventricle; without this process the blood could not properly aid in nourishing the fire that is in the heart. This is confirmed in the fact that one sees that animals without lungs have but one single ventricle in their hearts, and that children, who cannot use their lungs while locked within their mother's womb, have an opening through which blood flows from the *vena cava* into the left ventricle of the heart, as well as a tube through which the blood goes from the *vena arteriosa* to the aorta, without passing through the lungs. Now how does digestion take place in the stomach if the heart does not send heat there through the arteries, along with some of the more fluid parts of the blood that help dissolve the food placed there? And is it not easy to understand the action that changes the juices of the food into blood, if one considers that they are distilled, in passing and repassing through the heart, perhaps more than one or two hundred times a day? And need anything else be said to explain nutrition and the production of the body's 54 various humors, except that the force with which the blood, as it is being rarified, passes from the heart to the extremities of the arteries, makes some of its parts stop in those of the members where they are found and there take the place of others that they expel; and that, according to their location or shape, or the smallness of the pores they encounter, they tend to go some places rather than others, in just the same way, as anyone can see, as various sieves that, being variously perforated, help separate out different size grains from one another? And finally, what is most remarkable in all this is the generation of animal spirits that are like a very subtle wind, or better, like a very pure and lively flame that, rising continually in great abundance from the heart to the brain, and from there goes through the nerves into the muscles, and gives movement to all the members, without the need for imagining any other reason for the fact that the parts of blood which, being the most agitated and most penetrating, are the most likely to make up these spirits, go to the brain rather than elsewhere, except that the arteries that carry these parts of blood are those that go from the heart in the straightest line of all; for, according to the laws of mechanics, which are the same as the laws of nature, when several objects tend all together to move in a single direction, where there is not enough room for all of them, as the parts of blood leaving the left ventricle of the heart tend toward the brain, the weakest and least lively 55 must be pushed aside by the stronger which in this way arrive by themselves at the brain.

I have given a sufficient explanation for everything in the treatise that I had intended earlier to publish. And I went on to show of what sort the fabric of the nerves and muscles of the human body must be, so that the animal spirits within might have the force to move its members; thus one observes that heads, shortly after being severed, still move about, and bite the earth, even though they are no longer alive. I also showed what changes ought to take place in the brain to cause wakefulness, sleep, and dreams; how light, sounds, odors, tastes, heat, and all the other qualities of external objects can imprint various ideas through the medium of the senses; how hunger, thirst, and the other internal passions can also send their own ideas; what needs to be taken for the common sense, where these ideas are received, for the memory which conserves them, and for the imagination which can change them in various ways and make new ones out of them, and by the same means, distributing the animal spirits in the muscles, make the members of this body move in as many different ways, each appropriate to the objects presented to the senses and to the internal passions that are in the body, as our own bodies can move, without being lead to do so through the intervention of our will. This ought not seem strange to those who, cognizant of how many different automata or moving machines the ingenuity of men can devise, using only a very small number of parts, in comparison to the great multitude of bones, muscles, nerves, arteries, veins, and all the other parts which are in the body of each animal, will consider this body like a machine that, having been made by the hand of God, is incomparably better ordered and has within itself movements far more admirable than any of those machines that can be invented by men.

And I paused here particularly to show that, if there were such machines having the organs and the shape of a monkey or of some other nonrational animal, we would have no way of telling whether or not they were of the same nature as these animals; if instead they resembled our bodies and imitated so many of our actions as far as this is morally possible, we would always have two very certain means of telling that they were not, for all that, true men. The first means is that they would never use words or other signs, putting them together as we do in order to tell our thoughts to others. For one can well conceive of a machine being so made as to pour forth words, and even words appropriate to the corporeal actions that cause a change in its organs—as, when one touches it in a certain place, it asks what one wants to say to it, or it cries out that it has been injured, and the like—but it could never arrange its words differently so as to answer to the sense of all that is said in its presence, which is something even the most backward men can do. The second means is that, although they perform many tasks very well or perhaps can do them better than any of us, they inevitably fail in other tasks; by this means one would discover that they do not act through knowledge, but only through the disposition of their organs. For while reason is a universal instrument that can be of help in all sorts of circumstances, these organs require a particular disposition for each particular action; consequently, it is morally impossible for there to be enough different devices in a machine to make it act in all of life's situations in the same way as our reason makes us act.

For by these two means one can know the difference between men and beasts. For it is very remarkable that there are no men so backward and so stupid, excluding not even fools, who are unable to arrange various words and to put together discourse through which they make their thoughts understood; but, on the other hand, there is no other animal, perfect and well bred as it may be, than can do likewise. This is not due to the fact that they lack the organs for it, for magpies and parrots can utter words just as we can, and still they cannot speak as we can, that is, by giving evidence of the fact that they are thinking about what they are saying; although the deaf and dumb are deprived just as much as— or more than—animals of the organs which aid others in speaking, they are in the habit of inventing for themselves various signs through which they make themselves understood to those who are usually with them and have the leisure to learn their language. And this attests not merely to the fact that animals have less reason than men but that they have none at all. For one sees that not much of it is needed so as to be able to speak; given that one notices an inequality among animals of the same species, just as is the case among men, and that some are easier to train than others, one could not believe that a monkey or a parrot which is the most perfect of its species would not equal in this one of the most stupid of children, or at least a child with a disturbed brain, unless their soul were of an ut- terly different nature from our own. And one should not confuse words with natural movements that display the passions and can be imitated by machines as well as by animals; nor should we think, as did some ancients, that animals speak, although we do not understand their language; for if that were true, since they have many organs similar to our own, they could also make themselves understood by us just as they are by their peers. It is also remarkable that, although there are many animals that show more inventiveness than we do in some of their actions, one never- theless sees that they show none at all in many other actions; conse- quently, the fact that they do something better than we do does not prove that they have a mind; for were this the case, they would be more rational than any of us and would excel us in everything; but rather it proves that they do not have a mind, and that it is nature that acts in them, according to the disposition of their organs—just as one sees that a clock made only of wheels and springs can count the hours and measure time more accurately than we can with all our powers of reflective de- liberation.

After this, I described the rational soul and showed that it can in no way be drawn from the potentiality of matter, as can the other things I have spoken of, but that it ought expressly to be created; and how it is not enough for it to be lodged in the human body, like a pilot in his ship, unless perhaps to move its members, but it must be joined and united more closely to the body so as to have, in addition, feelings and appetites similar to our own, and thus to make up a true man. As to the rest, I elaborated here a little about the subject of the soul because it is of the greatest importance; for, after the error of those who deny the existence of God (which I believe I have sufficiently refuted), there is nothing that puts weak minds at a greater distance from the straight road of vir-

58

59

tue than imagining that the soul of animals is of the same nature as ours and that, as a consequence, we have no more to fear nor to hope for after this life than have flies or ants; whereas, while one understands how much they differ, one understands much better the reasons that prove that our soul is of a nature entirely independent of the body and, consequently, that it is not subject to die with it. Now, since one cannot see any other causes that might destroy it, one is naturally lead to judge from this that the soul is immortal.

60

Part Six

But it is now three years since I completed the treatise containing all these things, and I began to review it in order to put it into the hands of a publisher, when I learned that people to whom I defer, and whose authority over my actions cannot be less than that of my reason over my thoughts, had disapproved of a certain opinion in the realm of physics, published a short time before by someone else,[7] concerning which I do not wish to say that I was in agreement, but rather that I had found nothing in it, before their censuring of it, that I could imagine to be prejudicial either to religion or to the state, nor had I found anything that would have stopped me from writing it, had reason persuaded me of it; and this made me fear that I might likewise find among my opinions some in which I was mistaken, not withstanding the great care that I had always taken not to accept into my set of beliefs any new opinions for which I did not have very certain demonstrations and never to write anything that could turn to someone's disadvantage. This was sufficient to make me change the resolution I had made to publish them. For although the reasons for which I had earlier made the resolution were very strong, my inclination, always making me hate the business of writing books, immediately made me find enough other reasons to excuse me from it. And these reasons, both for and against, were such that not only do I have an interest in saying these things, but perhaps the public too has an interest in knowing them.

61

I had never made much of what came from my mind, and as long as I had reaped no other fruits from the method which I used, aside from my own satisfaction, in regard to certain problems that pertain to the

---

[7] Galileo Galilei (1564–1642), Italian astronomer, mathematician and physicist. His *Dialogue . . . on the Two Chief Systems of the World* (1632), in which he advanced the theory of the movement of the earth, occasioned the Inquisitors of the Holy Office to conduct a trial in Rome and to extort a retraction of that theory from Galileo. Descartes, who also advocated a theory of terrestrial motion, was not about to let Rome sin twice against philosophy. Cf. E. Gilson, *Discours de la méthode: texte et commentaire*, pp. 439–442.

speculative sciences or my attempt at governing my moral conduct by means of the reasons which the method taught me, I believed I was under no obligation to write anything. For as to moral conduct, each person agrees so much with his own opinion, that one could find as many reformers as heads, were it permitted for those other than the ones God has established as rulers over his peoples or to whom he has given sufficient grace and zeal to be prophets to try to change anything; although my speculations pleased me very much, I had believed the others also had their speculations which perhaps pleased them even more. But as soon as I had acquired some general notions in the area of physics, and, beginning to test them on various specific difficulties, I had noticed just how far they can lead and how much they differ from the principles that people have used up until the present, I believed I could not keep them hidden away without greatly sinning against the law that obliges us to procure as best we can the common good of all men. For these general notions show me that it is possible to arrive at knowledge that is very useful in life and that in place of the speculative philosophy taught in the Schools, one can find a practical one, by which, knowing the force 62 and the actions of fire, water, air, stars, the heavens, and all the other bodies that surround us, just as we understand the various skills of our craftsmen, we could, in the same way, use these objects for all the purposes for which they are appropriate, and thus make ourselves, as it were, masters and possessors of nature. This is desirable not only for the invention of an infinity of devices that would enable us to enjoy without pain the fruits of the earth and all the goods one finds in it, but also principally for the maintenance of health, which unquestionably is the first good and the foundation of all the other goods in this life; for even the mind depends so greatly upon the temperament and on the disposition of the organs of the body that, were it possible to find some means to make men generally more wise and competent than they have been up until now, I believe that one should look to medicine to find this means. It is true that the medicine currently practiced contains little of such usefulness; but without trying to ridicule it, I am sure that there is no one, not even among those in the medical profession who would not admit that everything we know is almost nothing in comparison to what remains to be known, and that we might rid ourselves of an infinity of maladies, both of body and mind, and even perhaps also the enfeeblement brought on by old age, were one to have a sufficient knowledge of their causes and of all the remedies that nature has provided us. For, desiring to spend my entire life searching for so needed a science, and having found a 63 road that seems to me such that, by following it, one ought infallibly to find that science, were it not for the fact that one is stopped either by the brevity of life or by a lack of experience, I judged there to be no

better remedy against these two impediments than thus to convey faith-
fully to the public what little I had found and to urge good minds to
try to advance beyond this, by contributing, each according to his inclination and ability, to the experiments one must conduct and also by
conveying to the public everything they learned, so that later inquirers,
beginning where their predecessors had left off, and thus, in joining together the lives and works of many, we might all together advance much
further than a single individual could on his own.

   Moreover, I noticed, in regard to experiments, that they become more
necessary as one becomes more advanced in knowledge. For in the beginning it is better to make use only of what presents itself to our senses
of its own accord and which we could not ignore, provided we reflect
just a little on it, than to search for unusual and contrived experiments.
The reason is that the most unusual ones often deceive one when one does
not know yet the causes of the most ordinary experiments, and that the
circumstances on which the unusual ones depend are almost always so
specific and minute that it is very difficult to observe them. But the
order I have held to has been the following. First, I tried to find in a
64   general way the principles or first causes of all that is or can be in the
world, but not considering anything to this end except God alone who
created the world, and not drawing these principles from any other
source but from certain seeds of truth that are in our souls. After this
I examined which ones were the first and most ordinary effects that could
be deduced from these causes; it seemed to me that I had thus found the
heavens, stars, an earth, and even, on the earth, water, air, fire, minerals,
and other things that are the most common of all and the simplest—and
hence the easiest to know. Then, when I wanted to descend to the more
particular ones, so many different ones were presented to me that I did
not believe it possible for the human mind to distinguish the forms or
species of bodies that are on the earth from an infinity of others that
could have been—had it been the will of God to have put them there—
or, as a consequence, to make them serviceable to us, unless one goes
ahead to causes through effects and makes use of many particular experiments. After this, passing my mind again over all the objects that
ever presented themselves to my senses, I dare say that I have never seen
anything that I could not explain with sufficient ease through the principles I have found. But it is also necessary for me to admit that the
power of nature is so ample and so vast, and that these principles are so
simple and so general, that I observe almost no specific effect without
65   my first knowing that it can be deduced in many different ways, and
that my greatest difficulty is ordinarily to find in which of these ways
the effect actually depends. For, to this end, I know of no other ex-
pedient except to search once more for some experiments that are such

that their outcomes are not the same, if it is in one of these ways rather than in another that one ought to explain the effect. As to the rest, I am now at the point where, it seems to me, I see quite well in what direction one must go in order to do the majority of the experiments that can serve this purpose; but I also see that they are of such a nature and in such a multitude that neither my hands nor my financial resources, even if I had a thousand times more than I had, would be sufficient for all of them; so that, according as I henceforth have the wherewithal to do more or less of them, I shall also more or less go forward in the knowledge of nature. I promised myself to make this known through the treatise I have written, and to show in it so clearly the utility that the public can gain from it; I urge all those who desire the general well being of men, that is, all those who really are virtuous and not through false pretenses or merely through reputation, to communicate those experiments that they have already done as well as to help me in the search for those that remain to be done.

But since then I have had other reasons that have made me change my mind and to believe that I really ought to continue to write about all the things I judged of any importance as soon as I discovered the truth with respect to them, and to take the same care as I would if I wanted to print them. I did this as much to have more of an occasion to examine them   66
carefully (since, without doubt one always looks more carefully at what one believes will be seen by many than at what one does only for oneself; and after the things that seemed to me to be true when I began to conceive them have appeared false to me when I have decided to put them on paper), as not to lose the occasion to benefit the public, if I am up to it, and so that, if my writings are worth anything, those who will have them after my death can use them in the way that would be most fitting; but that I should absolutely refuse to have them published during my lifetime, so that neither the hostilities and the controversies to which they might be subject, nor even such reputation as they might gain for me would give me any occasion for wasting the time I have intended to use for self-instruction. For although it is true that each man is obliged to see as best he can to the good of others, and that being useful to no one is actually to be worthless, still it is also true that our concern ought to extend further than the present, and that it is well to omit things that perhaps would yield a profit to those who are living, when it is one's purpose to do other things that yield even more profit to our posterity. And I really want people to understand that what little I have learned up until now is almost nothing in comparison to what I do not know and to what I do not despair of being able to know; for it is almost the same with those who little by little discover the truth in the sciences,   67
as it is with those who, upon beginning to become wealthy, have less

trouble in making large acquisitions than they had had before when they were poorer in making very small ones. Or one might well compare them to the leaders of an army whose forces regularly grow in proportion to their victories; they need more leadership to maintain themselves after the loss of a battle than they do after they have succeeded in taking cities or provinces. For one truly engages in battles when one tries to overcome all the difficulties and mistakes that keep us from arriving at the knowledge of the truth; and it is really a loss of a battle to accept a false opinion touching on a rather general and important matter, and it requires afterward much more skill to recover one's earlier position than to make great progress when one already has principles that are certain. For myself, if I have already found any truths in the sciences (and I hope the things contained in this volume will cause one to judge that I have found some), I can say that these are only the results and offshoots of five or six major difficulties that I have surmounted; and I count them as so many battles in which luck was on my side. I will not even be fearful of saying that I think I only needed to win two or three others like them and I shall have entirely achieved the completion of my plans, and that my age is not so advanced that, according to the usual course of nature, I might not still have enough leisure to bring this about. But I believe I am all the more obligated to manage well the time remaining to me, the more hope I have of being able to use it well; doubtless I had many opportunities to waste it, had I published the foundations of my physics. For although they are nearly all so evident that it is necessary only to understand them in order to believe them, and although there has never been one for which I did not believe I could give demonstrations, nevertheless, since it is impossible for them to be in agreement with all the various opinions of other men, I foresee that I would often be distracted by the hostilities these matters would engender.

One could say that these hostilities might be useful both to make me aware of my faults, as well as, were I to have anything worthwhile, to make in this way others more knowledgeable about it, and, since many can see more than one man alone, that, beginning right now to use my method, others might help me also with their inventions. But, although I acknowledge that I am extremely prone to err and that I almost never rely on the first thoughts that come to me, still the experience I have of the objections one might make against me stops me from hoping for any profit from all this. For I have already often met with the judgments both of those I took to be my friends, as well as of various others whom I took to be indifferent, and even of those too whose maliciousness and envy I knew would try very hard to discover what affection would hide from my friends. It is rare that anyone has raised an objec-

tion to me that I had not foreseen at all, unless it were very far removed   69
from my subject; thus I have almost never found any critic of my opinions who did not seem either less rigorous or less objective than myself.
And I have never observed that, through the method of disputation
practiced in the Schools, any truth was discovered that had until then
been unknown. For, while each person in the dispute tries to win, he is
more concerned with putting on a good show than with weighing the
arguments on both sides; and those who long have been strong advocates
are not, on that account, better judges later.

As to the utility that others would receive from the communication
of my thoughts, it cannot be so terribly great, given that I have not yet
taken them so far that there is not any need to add many things before
applying them to common use. And I believe I can say without conceit
that, if there is anyone who can do this, it ought to be me rather than
someone else: not that there cannot be in the world many minds incomparably greater than my own, but that one cannot conceive a thing so
well and make it one's own when one learns it from another as one can
when one discovers it for oneself. This is so true in this matter that, although I have often explained some of my opinions to people with great
minds who, while I spoke to them, seemed to understand these opinions
quite well, nevertheless, when they repeated them, I noticed that they
had almost always altered them in such a way that I could no longer
acknowledge them to be mine. At this time I am very happy to ask our   70
posterity never to believe the things people say came from me, unless I
myself have revealed them. And I am not at all surprised at the extravagances attributed to all those ancient philosophers whose writings we do
not have; on that account I do not judge that their thoughts were terribly
unreasonable, given that they were the greatest minds of their time, but
only that someone has given us a bad account of them. For one sees also
that their followers almost never surpassed them; and I am sure that the
most impassioned of those who now follow Aristotle would believe
themselves fortunate, were they to have as much knowledge of nature
as he had, even if a condition of this were that they could never have
any more knowledge. They are like ivy that tends to climb no higher
than the trees supporting it, and even which often tends downward again
after it has reached the top; for it seems to me also that those in the
Schools also tend downward again, that is, they make themselves somehow less learned than had they abstained from studying, who, being
unhappy with knowing only what is intelligibly explained in their author,
also desire to find the solutions to many difficulties about which he has
said nothing and about which he has never thought. Still, their manner
of philosophizing is very congenial to those who have only very medi-
ocre minds, for the obscurity of the distinctions and the principles they

use is the reason why they can speak as boldly of all those things as if they know them and why they can maintain everything they say against the subtlest and most capable, without there being any way of convincing them. In this they seem to me like a blind man, who, to fight without a handicap against someone who is sighted, makes his opponent go into the depths of a very dark cave, and can I say that they have an interest in my abstaining from publishing the principles of the philosophy I use; for my principles being very simple and very evident, I would, by publishing them, be doing almost the same as if were I to open some windows and make some light of day enter that cave where they have descended to fight. But even the greatest minds have no reason for wanting to know them; for if they want to know how to speak about everything and to acquire the reputation for being learned, they will achieve their goal more easily by contenting themselves with the appearance of truth, which can be found without much difficulty in all sorts of matters, rather than by seeking the truth, which can only be discovered little by little in a few matters and which, when it is a question of speaking about other matters, obliges one to confess frankly that he does not know them. But if they prefer the knowledge of a few truths to the vanity of appearing to be ignorant of nothing, as undoubtedly is preferable, and if they desire to follow a plan similar to my own, they do not, on this account, need me to say anything more except what I have already said in this discourse. For, if they are able to pass further than I have done, they are also all the more able to find for themselves everything that I think I have found. Given that, having examined everything in an orderly way, it is certain that what still remains for me to discover is in itself more difficult and more hidden than what I have found up to this time; and they would take much less pleasure in learning it from me than from themselves. Moreover, the habit they will acquire of seeking first the easy things, then passing gradually by degrees to more difficult ones, will serve them better than all my instructions could do. As for myself, I am convinced that, had someone taught me from my youth all the truths for which I have sought demonstrations, and had I had no difficulty in learning them, I might perhaps have never learned any other truths, and at least I would never have acquired the habit and faculty I think I have for finding new truths, to the extent I apply myself in searching for them. And, in a word, if there is in the world any task that cannot be finished by anyone but the person who began it, it is that on which I am now working.

It is true that, with respect to experiments that can help here, one man alone cannot suffice to do them all, but he cannot as profitably use hands other than his own, unless they be those of craftsmen, or of such people as he can pay and for whom the hope of gain (which is a very effective means) would make do precisely what he ordered them to do. For, as to volunteers, who, out of curiosity or a desire to learn, offer themselves in order perhaps to help him, aside from usually being high on promises and low on performance and from having grand ideas none of which will come to anything, they inevitably want to be paid by an explanation of various difficulties, or at least by compliments or by useless conversations which could not cost him so little of his time that it would not be a loss to him. And as to the experiments that others have already done, even when these people would desire to communicate them to him (what those who regard them as their secrets never do), they are for the most part composed of so many details and superfluities that it would be very hard to discern the truth in them; besides, one finds almost all of them to be so badly explained or even so false, since those who have done them force themselves to make them appear in conformity with their principles, that, had there been some experiments they might use, they cannot be worth the time one would have to spend in choosing them. Thus if there were in the world anyone whom one knows with certainty to be capable of finding the greatest and most beneficial things possible, and for this reason the other men fully exerted themselves to help him succeed in his plans, I do not see that they could do a thing for him except to make a donation toward the expense of the experiments he needs and, for the rest, to keep his leisure from being wasted by the importunity of anyone. But, although I do not so much presume of myself to want to promise anything out of the ordinary, nor do I feed on such vain thoughts, as to imagine that the public ought especially to be interested in my plans, still I do not have so base a soul that I wish to accept from anyone any favor that someone might think I did not deserve.

All these considerations taken together were the reason why, three years ago, I did not want to unveil the treatise I had on hand, and why I had made a resolution not to make public any other treatise during my lifetime, which was so general, or one on the basis of which one could understand the foundations of my physics. But since then there have been yet again two more reasons that obliged me to attach here certain specific essays and to give the public some accounting of my actions and my plans. The first is that, if I failed to include it before, many who knew of the intention I once had to have published certain writings could imagine that the reasons why I abstain from doing so were less to my credit than was actually the case. For granted I did not love glory excessively, or even, if I dare say, that I hated it inasmuch as I judge it contrary to tranquility which I esteem above all things, still too, I had never tried to hide my actions as if they were crimes, nor had I used a lot of precaution so as not to be known; this is so both because

I had believed it would do me harm as well as because it would have given me a certain disquiet that again would have been contrary to the perfect repose of the mind I seek. Since, being always thus indifferent as to whether I am known or not, I could not prevent my acquiring a certain type of reputation, I believed I ought to do my best at least to save myself from having a bad one. The other reason that obliged me to write **75** this is that, seeing more and more everyday the delay that my plan of self-instruction is suffering, because of the infinity of experiments I need to make and that it is impossible to carry out without the help of someone else, although I do not flatter myself into hoping that the public will become greatly involved in my affairs, still I do not wish so much to fail myself as to give cause to those who survive me to reproach me someday for having been able to leave them many things that were much better than I had done, had I not overly neglected to make them understand how they could contribute to my plans.

And I have thought that it was easy to choose certain matters that, without being subject to a lot of controversy or obliging me to declare more of my principles than I desired, would allow me to show quite clearly what I can or cannot do in the sciences. As to this I cannot say whether I have been successful, and I do not want to prejudice the judgments of anyone in speaking for myself about my writings; but I would be very happy were a person to examine these, and, so that a person might have more of an opportunity to do this, I am asking all who have objections to make to take the trouble to send them to my publisher and, being advised about them by the publisher, I shall try to publish my reply at the same time as the objections; and by this means, seeing both of them together, readers will more easily judge the truth of the matter. For I do not promise to make long replies, but only to admit my errors **76** very candidly, if I know of any, or, if I cannot find any, to say simply what I believed was needed for the defense of what I have written—without adding an explanation of anything new—so that I am not engaging one objector after another in an endless procession.

And if any of those matters about which I spoke at the beginning of the *Dioptrics* and the *Meteorology* be shocking at first glance, because I call them "suppositions" and seem unwilling to prove them, I entreat the reader to have the patience to read the whole thing with attention; I hope he will find it satisfactory. For it seems to me that the reasons follow each other in such a way that, just as the last are demonstrated by the first—which are their causes—so these first are reciprocally demonstrated through the last—which are their effects. And one ought not to imagine that I am here committing the error logicians call a vicious circle; for, experience making the majority of these effects very certain, the causes from which I deduce these effects serve not so much to prove the effects as to explain them—but, on the other hand, the causes are what is proven by the effects. And I have called them "suppositions" only so that a person understands that I believe I can deduce them from the first truths that I have explained above. But I have expressly not wanted to do so, in order to prevent certain minds that imagine they know in one day all that someone else has thought about for twenty years as soon as he has said but two or three words to them, and who are the more subject to error and less capable of truth—the more penetrating and   77 lively they are—from being able there to take the occasion to build some extravagant philosophy on what they believe are my principles—and to prevent this from being attributed to me. For as to the opinions that are entirely my own, I do not excuse them for being new, given that, were one to consider well the reasons, I am certain that one would find them so simple and so in conformity with common sense that they seem less extraordinary and less strange than any others one could have on the same subjects. And I do not crow about being the first discoverer of any of them, but rather that I have never received them either because they have been said by others, or because they have not been, but only because reason persuades me of them.

If craftsmen cannot immediately execute the invention explained in the *Dioptrics*, I do not believe one can say, on that account, that it is bad; for, since skill and habitual disposition are needed to make and maintain the machines I have described, without any detail being overlooked, I would be no less astonished were they to succeed on the first try than were someone able to learn in one day to play the lute with distinction simply because one has been given a good musical score. And if I write in French, the language of my country, rather than in Latin, the language of my teachers, it is because I hope that those who use only their natural reason in its purity will judge my opinions, rather than those who believe only in old books. And as to those who combine good sense with study, to whom alone I submit as my judges, they will not, I am   78 certain, be so partial to Latin that they refuse to understand my reasons because I explain them in a vernacular language.

As to the rest, I do not want to speak here in specifics about the progress I hope to make in the future in the sciences, or to make any promise to the public that I am not certain of keeping; but I shall say simply that I have resolved to spend my remaining lifetime only in trying to acquire a knowledge of nature which is such that one could deduce from it rules for medicine that are more certain than those in use at present, and that my inclination puts me at such a great distance from any other sort of plan—principally those that can be useful to some person only while harmful to others—that if circumstances force me to busy myself with them, I do not believe I could succeed. On this I here declare that I know very well that it cannot help make me a man of stature in the eyes of the world, but for the matter I have no desire to be one; and I shall always try to be more obliged to those by whose favor I shall enjoy my leisure without obstacle than to those who might offer me the most honorable positions on earth.

END

Meditations
on
First Philosophy

NOTE ON THE TRANSLATION

The translation is based entirely on the Latin version of the *Meditations* found in volume seven of the Adam and Tannery edition of Descartes' works. It has been argued by Baillet, Descartes' early biographer, that the French "translation" by de Luynes is superior to the Latin version because it contains many additions and clarifications made by Descartes himself. However, I have not used the French version because it contains inconsistencies and shifts that muddle more than clarify the original Latin text. The numbers found in the margins of the present translation refer to the page numbers of the Latin text in the Adam and Tannery edition.

In one instance, I found that the Latin text did not square with Descartes' clear intention (p. 81 l. 25). The portion in brackets conveys my suggestion as to Descartes' actual intention in the passage.

D.A.C.

To the Wisest and Most Distinguished Men,
the Dean and Doctors of the Faculty of Sacred Theology of Paris
René Descartes Sends Greetings

Such a righteous cause impels me to offer this work to you (and I am confident that you too will regard it as so righteous as to take up its defense, once you have understood the plan of my undertaking) that there is here no better means of commending it than to state briefly what I have sought to achieve in this work.

I have always thought that two questions—that of God and that of the soul—are chief among those that ought to be demonstrated by the aid of philosophy rather than of theology. For although it suffices for believers like ourselves to believe by faith that the soul does not die with the body and that God exists, certainly no unbeliever seems capable of being persuaded of any religion or even any moral virtue, unless these two are first proven to him by natural reason. And since in this life there are often more rewards for vices than for virtues, few would prefer what is right to what is useful, if they neither feared God nor hoped for an afterlife. And although it is utterly true that God's existence is to be believed in because it is taught in the Holy Scriptures, and, on the other hand, that the Holy Scriptures are to be believed because they have God as their source (because, since faith is a gift from God, the very same one who gives the grace that is necessary for believing the rest, can also give us the grace to believe that he exists); nonetheless, this cannot be proposed to unbelievers becaues they would judge it to be a circle. And truly I have noticed that you, along with all other theologians, affirm not only that the existence of God can be proven by natural reason, but also that one may infer from the Holy Scriptures that the knowledge of him is much easier than the manifold knowledge that we have of created things, and is so utterly easy that those without this knowledge are worthy of blame. For this is clear from Wisdom, Chapter 13 where it is said: "They are not to be excused, for if their capacity for knowing were so great that they could appraise the world, how is it that they did not find the Lord of it even more easily?" And in Romans, chapter 1, it is said that they are "without excuse." And again in the same text we seem to be warned by these words: "What is known of God is manifest in them": everything that can be known about God can be made manifest by reason drawn from a source none other than our own mind. For this reason I have not thought it unbecoming for me to inquire how it is that this is the case, and by what path God may be known more easily and with greater certainty than the things of this world.

And as to the soul: although many have regarded its nature as incapable of easy inquiry, and some have gone so far as to say that human reasoning convinces them that the soul dies with the body, and that the contrary is to be held on faith alone; nevertheless, because the Lateran Council under Leo X, in Session 8, condemned these people and explicitly enjoined Christian philosophers to refute their arguments and to use all their abilities to make the truth known, I too have not hesitated to go forward with this.

Moreover, I know that there are many irreligious people who refuse to believe that God exists and that the soul is distinct from the body, for no other reason than that they say that these two doctrines have up to this time not been able to be proven by anybody. Although I am by no means in agreement with these people (on the contrary, I believe that nearly all the arguments which have been brought to bear on these questions by great men have the force of a demonstration, when they are adequately understood, and I am convinced that hardly any arguments could be given that were not previously discovered by others); nevertheless, I judge that there is no greater purpose to fulfill in philosophy than to seek out the best of all these arguments and to exhibit them so carefully and accurately that henceforth all will take them to be true demonstrations. And, finally, because I was very strongly urged to do this by several people, to whom it is known that I practiced a method for solving certain problems in the sciences—not a new one, because nothing is more ancient than the truth, but which I often seemed to use with success in other areas. Thus I believed that I should attempt something on this subject.

Now all that I have been able to accomplish is contained in this treatise. Not that I have begun to gather together in it all the various proofs that can be brought to bear on these matters; that does not seem necessary, except where no proof is sufficiently certain. Rather, I have sought the primary and principal arguments, so that I now dare to propose these as most certain and evident demonstrations. Moreover, these arguments are such that I believe that there is no other way by which human ingenuity can find better ones. For the urgency of the cause, as well as the glory of God to which all this is referred, compel me to speak here on my own behalf somewhat more freely than is my custom. But although I believe my arguments to be certain and obvious, still I am not therefore convinced that they have been accommodated to everybody's power of comprehension. Just as in geometry there are many proofs written by Archimedes, Apollonius, Pappus, and others, that, although they are taken to be obvious and certain (because they manifestly contain nothing which, considered by itself, is not quite easily known, and nothing in which what follows does not square exactly with what precedes), are nevertheless quite lengthy and require a very attentive reader, so that they are understood by only a few; just so, although I believe that the proofs I use here equal and even surpass in certitude and obviousness the demonstrations of geometry, I am nevertheless fearful that they would be inadequately perceived by many people, both because they are quite lengthy, one thing depending on another, and also because they particularly demand a mind quite free from prejudices—a mind that can easily withdraw itself from commerce with the senses. Certainly one is less apt to find people competent to study metaphysics than to study geometry. Moreover, there is a difference in that in geometry everyone is convinced that nothing is customarily written without there being a certain demonstration for it, so that the inexperienced err on the side of assenting to what is false, wanting as they do to give the appearance of understanding it, more often than of denying what is true. But it is the reverse in philosophy: since nothing is believed concerning which there cannot be a dispute regarding at least one part, few look for truth, and many more, eager to have a reputation for profundity, dare to challenge whatever is the best.

And therefore, however forceful my proofs might be, nevertheless—because they belong to philosophy—I do not expect that what I have accomplished through them will be very significant unless you assist me with your patronage. So great an esteem for your Faculty resides in the minds of all, and so authoritative is the name of the Sorbonne, that, not only in matters of faith, but also in humanistic philosophy, no society, other than the Holy Councils, has been deferred to so much as yours. Nowhere is there thought to be greater sharpness of wit and solid knowledge, or greater integrity and wisdom in delivering judgment. I do not doubt that, if you should deign to pay heed to this work, you would first correct it (for mindful that I am not only human but also ignorant, I do not assert that there are no errors in it); next, that what is lacking may be added, or what is not quite complete may be perfected, or what is in need of further discussion may be more fully laid out, either by yourselves or at least by me, after I have been made aware of it by you; and finally, that, after the arguments contained in it (by which we prove that God exists and that the mind is distinct from the body) will have been brought to that level of sharpness to which I am confident they can be brought, so that these arguments should come to be regarded as the most carefully prepared of demonstrations, should you wish to declare these and attest to them in public. I do not doubt, I say, that, if this should come to pass, all the errors that have ever been entertained regarding these questions will in a short time be erased from the minds of men. For the truth itself easily brings it about that the remaining men of intelligence and learning subscribe to your judgment; and your authority will bring it about that the atheists, who are more accustomed to being dilettantes than brilliant or learned men, shall put aside their spirit of contrariness, and also that perhaps they will defend the arguments which they will know are taken to be demonstrations by men of intelligence, lest they seem not to understand them. And finally, all the others will easily believe in so many testimonies, and there will be no one who would dare call into doubt either the existence of God or the real distinction of the soul from the body. Just how great the usefulness of this thing is, you yourselves can best of all be the judge, in virtue of your singular wisdom; nor does it behoove me to commend the cause of God and religion to you at any greater length, you who have always been the greatest pillar of the Catholic Church.

I have already touched lightly on the question of God and the human
mind in my *Discourse on the Method of Rightly Conducting the Reason
and Searching for Truth in the Sciences,* published in French in 1637,
not for the purpose of giving them an exhaustive treatment there, but
only, by sampling opinion, to learn from the judgment of readers how
these matters should be subsequently treated. For they seemed to me to
be so important that I judged that they ought to be dealt with more
than once. And the path I follow in order to explicate these questions is
so little trodden and so far removed from everyday use that I did not
believe it useful to profess at greater length in a work written in French
and read indiscriminately by all sorts of people, lest weaker minds be in
a position to believe that they too are to set out on this path.

In the *Discourse* I requested all of those to whom there might occur
something in my writings worthy of refutation that they would see fit
to warn me of it, and none of the objections, among the questions I received, were worth noting, except two; and I will respond to them here
in a few words before I undertake a more careful explanation of them.

The first is that, from the fact that the human mind, being turned in
on itself, perceives that it is only a thinking thing, it does not follow that
its nature or essence consists only in its being a thinking thing such that
the word "only" would exclude everything else which also could perhaps be said to pertain to the nature of the soul. To this objection I
answer that I also had no intention in that treatise of excluding those
things from the order of the truth of the matter (that is to say, I was
not dealing with that order then), but only as far as applies to the order
of my perception; thus my meaning was that I am plainly aware of
nothing which I know to pertain to my essence, beyond the fact that I
am a thinking thing, that is, a thing having within itself the faculty of
thinking. But in what ensues I will show how it follows from the fact
that I know nothing else that pertains to my essence that nothing else at
all really pertains to it.

The second objection is that it does not follow from the fact that I
have in myself the idea of a thing more perfect than I, that this idea is
more perfect than I, and much less that what is represented by this idea
exists. But I answer here that there is an equivocation in the word "idea,"
for it can be taken either materially, for an operation of the intellect (in
this sense one cannot say that this idea is more perfect than I), or else
objectively, for the thing represented by means of this operation. This
thing, although it is not presumed to exist outside the intellect, never-

7

8

theless can be more perfect than I by reason of its essence. But how, from the mere fact that there is in me an idea of something more perfect than I, it follows that this thing truly exists, will be manifested in detail in the ensuing remarks.

9   Moreover, I have seen two rather long pieces of writing, but in these it was not so much my arguments regarding these things that were attacked by arguments drawn from atheist commonplaces, but the conclusions. And because arguments of this type have no power over those who understand my premises, and because the judgments of most people are so preposterous and silly that they are more likely to be persuaded by the first opinions to come along, however false and alien to reason they may be, than by a true and firm, though subsequently received refutation of them, I do not wish to respond to these works here, lest I have to mention them first. Without going into particulars, I will say only that those considerations that the atheists crudely keep bringing up in order to assail the existence of God always depend on the fact that human sentiments are attributed to God, or that such power and wisdom are claimed for our minds that we begin to determine and grasp what God can and ought to do. And these considerations will cause us no difficulty, provided we only remember that our minds must be considered finite, and that God is incomprehensible and infinite.

But now, after having once and for all put to the test the judgments of men, I here again approach these same questions regarding God and the human mind, and at the same time treat the beginnings of the whole of first philosophy, but in such a way that I have no expectation of approval from the vulgar and no wide audience of readers. Rather, I am an author to none who read these things but those who seriously meditate with me, who have the ability and the desire to withdraw their mind from the senses and at the same time from all prejudices. Such people I know all too well to be few and far between. As to those who do not take care to comprehend the order and series of my reasons but eagerly dispute over single conclusions by themselves, as is the custom for many—

10   those, I say, will derive little benefit from a reading of this treatise; and although perhaps they might find an occasion for quibbling in many spots, still it is not an easy matter for them to raise an objection that is either compelling or worthy of response.

But because I do not promise others that I will satisfy them in all things the first time around, and I do not claim for myself so much that I believe that I can foresee everything that will seem difficult to someone, I will first narrate these very thoughts in the Meditations, by means of which I seem to have arrived at a certain and evident knowledge of the truth, so that I examine whether I can persuade others by the same arguments by which I have been persuaded. Then I will reply to the objections of several naturally gifted and very well-schooled gentlemen, to whom these Meditations have been sent for examination before they were sent to press. For their objections were so many and varied that I have dared to hope that nothing will easily come to mind for any other people—at least nothing of importance—which has not yet already been touched upon by these gentlemen. And thus I also ask the readers not to form a judgment regarding the Meditations before they have deigned to read these objections and the replies I have made.

12   In the First Meditation the reasons are given why we can doubt all things, especially material things, as long, of course, as we have no other foundations for the sciences than the ones which we have had up until now. Although it is not readily apparent that this doubt is useful, still it is the more so in that it frees us of all prejudices and paves an easy path for leading the mind away from the senses. Finally, it makes it impossible for us to doubt further those things that we shall discover to be true.

   In the Second Meditation the mind, by using the freedom peculiar to it, supposes the nonexistence of all those things about whose existence it can have the least doubt, and judges that such a supposition is impossible unless the mind exists during that time. This is also of the greatest utility, since by this means it easily distinguishes what things pertain to it—that is, to an intellectual nature—and what things pertain to a body. But because perhaps many people expected to see proofs for the immortality of the
13   soul here, they are to be put on notice that I believe that I have begun to write in this work only what I have carefully demonstrated. Therefore I was unable to follow any order but the one in use among geometricians, which is to state first everything on which the proposition in question depends, before concluding anything regarding that proposition. But what is required principally and primarily for knowing that the soul is immortal is that we formulate as clear a concept of the soul as possible, patently distinct from any concept of a body. This is what has been done here. Moroeover, it is also required that we know that everything that we clearly and distinctly understand is true, in precisely the manner in which we understand them. This could not have been proven before the Fourth Meditation. Moreover, one must have a distinct concept of corporeal nature, which is formulated partly in the Second Meditation, and partly in the Fifth and Sixth. But from all this one ought to conclude that all things clearly and distinctly conceived as diverse substances—as, for example, mind and body are conceived—are indeed substances that are really distinct from one another. This is arrived at in the Sixth Meditation. The same thing is confirmed there from the fact that a body cannot be understood to be anything but divisible, whereas the mind cannot be understood to be anything but indivisible, for we cannot conceive of half a soul, as we can in the case of any body, however small. This is so much the case that the natures of mind and body are acknowledged to be not only diverse but even, in a manner of speaking, to be the contraries of one another. I have not written further on the matter in this treatise, both because these considerations suffice for showing that the

annihilation of the mind does not follow from the corrupt state of the
body—and therefore for giving mortal men hope in an afterlife—and also
because the premises from which the immortality of the mind can be
proven depend upon an account of the whole of physics: first, we must   **14**
know that absolutely all substances, that is, everything that needs to be
created by God in order to exist, are by their very nature incorruptible,
nor can they ever cease to be, unless by this very same God the denial
of whose concurrence to them would reduce them to nothing. Second,
we must notice that a body, taken in a general sense, is a substance and
that it therefore can never perish. But the human body, insofar as it differs from other bodies, is composed only of a certain configuration of
members, together with other accidents of the same sort. But the human
mind is not composed of any such accidents, but is a pure substance. For
although all its accidents may undergo change—so that it understands
some things, wills others, senses others, etc.—the mind does not for this
reason become something else. But the human body does become something else, by the mere fact that the shape of some of its parts changes.
From these considerations it follows that a body can easily cease to be,
whereas the mind by its nature is immortal.

In the Third Meditation I have explained at sufficient length, it seems
to me, my principal argument for proving the existence of God. Be that
as it may, I sought to use no comparisons drawn from corporeal things,
in order to draw the souls of the readers as far as possible from the senses.
Thus obscurities may perhaps have remained; but these, I hope, will later
be easily removed in my Replies to the Objections. One such obscurity
is this: how can the idea that is in us of a supremely perfect being have
so much objective reality that it must be from a supremely perfect cause?
This is illustrated in the Replies by a comparison with a very ingenious
machine, the idea of which is in the mind of the artisan. For, just as the
objective ingeniousness of this idea must have some cause (say, the
knowledge possessed by the artisan or by someone else from whom he
has received this knowledge), just so, the idea of God, which is in us,   **15**
must have God himself for its cause.

In the Fourth Meditation I prove that all that we clearly and distinctly
perceive is true; and, at the same time, I explain wherein the nature of
falsity consists. These things necessarily ought to be known, as much
for reinforcing what has just been said as for understanding what remains. But it should be borne in mind that in that Meditation I have not
discussed sin—that is, an error committed in the pursuit of good and
evil—but only that error which occurs in the discernment of the true and
the false. Nor do I examine those matters pertaining to religious faith
or to the conduct of life, but only those speculative truths that are known
by the aid of the light of nature.

In the Fifth Meditation in addition to explaining corporeal nature in general, I also demonstrate the existence of God by means of a new proof. But again some difficulties arise. These are later resolved in my Replies to the Objections. Finally, I show how it is true that the certitude of geometrical demonstrations depends upon a knowledge of God.

Finally, in the Sixth Meditation I distinguish understanding from imagination; I describe the marks of this distinction. I prove that the mind is really distinct from the body (although I show that the mind is so closely joined to the body, that it forms one thing with the body.) All the errors commonly arising from the senses are presented, along with all the ways in which these errors can be avoided. Finally, I append all the grounds on the basis of which the existence of material things can be inferred. Not that I believe these grounds to be very useful for proving what they prove—namely that there really is a world, that men have bodies, and other such things, concerning which no one of sound mind has ever seriously doubted—but because, by considering these grounds, I recognize that they are neither so firm nor so evident as are the grounds by means of which we arrive at a knowledge of our mind and of God, insofar as these latter grounds are, of all that can be known by the human mind, the most certain and the most evident. Proving this one thing was the goal of these Meditations. For this reason I pass over these other questions, which also will be given treatment in the Meditations at the appropriate occasion.

Meditations
on
First Philosophy
In Which
The Existence of God
And The Distinction of the Soul from the Body
Are Demonstrated

on
First Philosophy
In Which
The Existence of God
And The Distinction of the Soul from the Body
Are Demonstrated

Meditation One: Concerning Those Things That Can Be Called into Doubt

Several years have now passed since I first realized how many were the false opinions that in my youth I took to be true, and thus how doubtful were all the things that I subsequently built upon these opinions. From the time I became aware of this, I realized that for once I had to raze everything in my life, down to the very bottom, so as to begin again from the first foundations, if I wanted to establish anything firm and lasting in the sciences. But the task seemed so enormous that I waited for a point in my life that was so ripe that no more suitable a time for laying hold of these disciplines would come to pass. For this reason, I have delayed so long that I would be at fault were I to waste on deliberation the time that is left for action. Therefore, now that I have freed my mind from all cares, and I have secured for myself some leisurely and carefree time, I withdraw in solitude. I will, in short, apply myself earnestly and openly to the general destruction of my former opinions. 18

Yet to this end it will not be necessary that I show that all my opinions are false, which perhaps I could never accomplish anyway. But because reason now persuades me that I should withhold my assent no less carefully from things which are not plainly certain and indubitable than I would to what is patently false, it will be sufficient justification for rejecting them all, if I find a reason for doubting even the least of them. Nor therefore need one survey each opinion one after the other, a task of endless proportion. Rather—because undermining the foundations will cause whatever has been built upon them to fall down of its own accord— I will at once attack those principles which supported everything that I once believed.

Whatever I had admitted until now as most true I took in either from the senses or through the senses; however, I noticed that they sometimes deceived me. And it is a mark of prudence never to trust wholly in those things which have once deceived us.

But perhaps, although the senses sometimes deceive us when it is a question of very small and distant things, still there are many other matters which one certainly cannot doubt, although they are derived from the very same senses: that I am sitting here before the fireplace wearing my dressing gown, that I feel this sheet of paper in my hands, and so on. But how could one deny that these hands and that my whole body exist? Unless perhaps I should compare myself to insane people whose brains are so impaired by a stubborn vapor from a black bile that they continually insist that they are kings when they are in utter poverty, or that they are wearing purple robes when they are naked, or that they have a head made of clay, or that they are gourds, or that they are made of glass. But they are all demented, and I would appear no less demented if I were to take their conduct as a model for myself.

All of this would be well and good, were I not a man who is accustomed to sleeping at night, and to undergoing in my sleep the very same things—or now and then even less likely ones—as do these insane people when they are awake. How often has my evening slumber persuaded me of such customary things as these: that I am here, clothed in my dressing gown, seated at the fireplace, when in fact I am lying undressed between the blankets! But right now I certainly am gazing upon this piece of paper with eyes wide awake. This head which I am moving is not heavy with sleep. I extend this hand consciously and deliberately and I feel it. These things would not be so distinct for one who is asleep. But this all seems as if I do not recall having been deceived by similar thoughts on other occasions in my dreams. As I consider these cases more intently, I see so plainly that there are no definite signs to distinguish being awake from being asleep that I am quite astonished, and this astonishment almost convinces me that I am sleeping.

Let us say, then, for the sake of argument, that we are sleeping and that such particulars as these are not true: that we open our eyes, move our heads, extend our hands. Perhaps we do not even have these hands, or any such body at all. Nevertheless, it really must be admitted that things seen in sleep are, as it were, like painted images, which could have been produced only in the likeness of true things. Therefore at least these general things (eyes, head, hands, the whole body) are not imaginary things, but are true and exist. For indeed when painters wish to represent sirens and satyrs by means of bizarre and unusual forms, they surely cannot ascribe utterly new natures to these creatures. Rather, they simply intermingle the members of various animals. And even if they concoct something so utterly novel that its likes have never been seen before (being utterly fictitious and false), certainly at the very minimum the colors from which the painters compose the thing ought to be true. And for the same reason, although even these general things (eyes, head, hands, and the like) can be imaginary, still one must necessarily admit that at least other things that are even more simple and universal are true, from which, as from true colors, all these things—be they true or false—which in our thought are images of things, are constructed.

To this class seems to belong corporeal nature in general, together with its extension; likewise the shape of extended things, their quantity or size, their number; as well as the place where they exist, the time of their duration, and other such things.

Hence perhaps we do not conclude improperly that physics, astronomy, medicine, and all the other disciplines that are dependent upon the consideration of composite things are all doubtful. But arithmetic, geometry, and other such disciplines—which treat of nothing but the simplest and most general things and which are indifferent as to whether these things do or do not exist—contain something certain and indubitable. For whether I be awake or asleep, two plus three makes five, and a square does not have more than four sides; nor does it seem possible that such obvious truths can fall under the suspicion of falsity.

All the same, a certain opinion of long standing has been fixed in my mind, namely that there exists a God who is able to do anything and by whom I, such as I am, have been created. How do I know that he did not bring it about that there be no earth at all, no heavens, no extended thing, no figure, no size, no place, and yet all these things should seem to me to exist precisely as they appear to do now? Moreover—as I judge that others sometimes make mistakes in matters that they believe they know most perfectly—how do I know that I am not deceived every time I add two and three or count the sides of a square or perform an even simpler operation, if such can be imagined? But perhaps God has not willed that I be thus deceived, for it is said that he is supremely good. Nonetheless, if it were repugnant to his goodness that he should have created me such that I be deceived all the time, it would seem, from this same consideration, to be foreign to him to permit me to be deceived occasionally. But we cannot make this last assertion.

Perhaps there are some who would rather deny such a powerful God, than believe that all other matters are uncertain. Let us not put these people off just yet; rather, let us grant that everything said here about God is fictitious. Now they suppose that I came to be what I am either by fate or by chance or by a continuous series of events or by some other way. But because being deceived and being mistaken seem to be imperfections, the less powerful they take the author of my being to be, the more probable it will be that I would be so imperfect as to be deceived perpetually. I have nothing to say in response to these arguments. At length I am forced to admit that there is nothing, among the things I once believed to be true, which it is not permissible to doubt—not for

21

reasons of frivolity or a lack of forethought, but because of valid and considered arguments. Thus I must carefully withhold assent no less from these things than from the patently false, if I wish to find anything certain.

But it is not enough simply to have made a note of this; I must take care to keep it before my mind. For long-standing opinions keep coming back again and again, almost against my will; they seize upon my credulity, as if it were bound over to them by long use and the claims of intimacy. Nor will I get out of the habit of assenting to them and believing in them, so long as I take them to be exactly what they are, namely, in some respects doubtful as by now is obvious, but nevertheless highly probable, so that it is much more consonant with reason to believe them than to deny them. Hence, it seems to me, I would do well to turn my will in the opposite direction, to deceive myself and pretend for a considerable period that they are wholly false and imaginary, until finally, as if with equal weight of prejudice[1] on both sides, no bad habit should turn my judgment from the correct perception of things. For indeed I know that no danger or error will follow and that it is impossible for me to indulge in too much distrust, since I now am concentrating only on knowledge, not on action.

Thus I will suppose not a supremely good God, the source of truth, but rather an evil genius, as clever and deceitful as he is powerful, who has directed his entire effort to misleading me. I will regard the heavens, the air, the earth, colors, shapes, sounds, and all external things as nothing but the deceptive games of my dreams, with which he lays snares for my credulity. I will regard myself as having no hands, no eyes, no flesh, no blood, no senses, but as nevertheless falsely believing that I possess all these things. I will remain resolutely fixed in this meditation, and, even if it be out of my power to know anything true, certainly it is within my power to take care resolutely to withhold my assent to what is false, lest this deceiver, powerful and clever as he is, have an effect on me. But this undertaking is arduous, and laziness brings me back to my customary way of living. I am not unlike a prisoner who might enjoy an imaginary freedom in his sleep. When he later begins to suspect that he is sleeping, he fears being awakened and conspires slowly with these pleasant illusions. In just this way, I spontaneously fall back into my old beliefs, and dread being awakened, lest the toilsome wakefulness which follows upon a peaceful rest, have to be spent thenceforward not in the light but among the inextricable shadows of the difficulties now brought forward.

---

[1] A "prejudice" is a prejudgment, that is, an adjudication of an issue without having first reviewed the appropriate evidence.

Meditation Two: Concerning the Nature of the Human Mind: That the Mind is More Known Than the Body

Yesterday's meditation filled my mind with so many doubts that I can no longer forget about them—nor yet do I see how they are to be resolved. But, as if I had suddenly fallen into a deep whirlpool, I am so disturbed that I can neither touch my foot to the bottom, nor swim up to the top. Nevertheless I will work my way up, and I will follow the same path I took yesterday, putting aside everything which admits of the least doubt, as if I had discovered it to be absolutely false. I will go forward until I know something certain—or, if nothing else, until I at least know for certain that nothing is certain. Archimedes sought only a firm and immovable point in order to move the entire earth from one place to another. Surely great things are to be hoped for if I am lucky enough to find at least one thing that is certain and indubitable.

Therefore I will suppose that all I see is false. I will believe that none of those things that my deceitful memory brings before my eyes ever existed. I thus have no senses: body, shape, extension, movement, and place are all figments of my imagination. What then will count as true? Perhaps only this one thing: that nothing is certain.

But on what grounds do I know that there is nothing over and above all those which I have just reviewed, concerning which there is not even the least cause for doubt? Is there not a God (or whatever name I might call him) who instills these thoughts in me? But why should I think that, since perhaps I myself could be the author of these things? Therefore am I not at least something? But I have already denied that I have any senses and any body. Still, I hesitate; for what follows from that? Am I so tied to the body and to the senses that I cannot exist without them? But I have persuaded myself that there is nothing at all in the world: no heaven, no earth, no minds, no bodies. Is it not then true that I do not exist? But certainly I should exist, if I were to persuade myself of something. But there is a deceiver (I know not who he is) powerful and sly in the highest degree, who is always purposely deceiving me. Then there is no doubt that I exist, if he deceives me. And deceive me as he will, he can never bring it about that I am nothing so long as I shall think that I am something. Thus it must be granted that, after weighing everything carefully and sufficiently, one must come to the considered judgment that the statement "I am, I exist" is necessarily true every time it is uttered by me or conceived in my mind.

But I do not yet understand well enough who I am—I, who now necessarily exist. And from this point on, I must take care lest I imprudently substitute something else in place of myself; and thus be mistaken even in that knowledge which I claim to be the most certain and evident of

24

25

all. To this end, I shall meditate once more on what I once believed myself to be before having embarked upon these deliberations. For this reason, then, I will set aside whatever can be refuted even to a slight degree by the arguments brought forward, so that at length there shall remain precisely nothing but what is certain and unshaken.

What therefore did I formerly think I was? A man, of course. But what is a man? Might I not say a rational animal? No, because then one would have to inquire what an "animal" is and what "rational" means. And then from only one question we slide into many more difficult ones. Nor do I now have enough free time that I want to waste it on subtleties of this sort. But rather here I pay attention to what spontaneously and at nature's lead came into my thought beforehand whenever I pondered what I was. Namely, it occurred to me first that I have a face, hands, arms, and this entire mechanism of bodily members, the very same as are discerned in a corpse—which I referred to by the name "body." It also occurred to me that I eat, walk, feel and think; these actions I used to assign to the soul as their cause. But what this soul was I either did not think about or I imagined it was something terribly insubstantial—after the fashion of a wind, fire, or ether—which has been poured into my coarser parts. I truly was not in doubt regarding the body; rather I believed that I distinctly knew its nature, which, were I perhaps tempted to describe it such as I mentally conceived it, I would explain it thus: by "body," I understand all that is suitable for being bounded by some shape, for being enclosed in some place, and thus for filling up space, so that it excludes every other body from that space; for being perceived by touch, sight, hearing, taste, or smell; for being moved in several ways, not surely by itself, but by whatever else that touches it. For I judged that the power of self-motion, and likewise of sensing or of thinking, in no way pertains to the nature of the body. Nonetheless, I used to marvel especially that such faculties were found in certain bodies.

But now what am I, when I suppose that some deceiver—omnipotent and, if I may be allowed to say it, malicious—takes all the pains he can in order to deceive me? Can I not affirm that I possess at least a small measure of all those traits which I already have said pertain to the nature of the body? I pay attention, I think, I deliberate—but nothing happens. I am wearied of repeating this in vain. But which of these am I to ascribe to the soul? How about eating or walking? These are surely nothing but illusions, because I do not have a body. How about sensing? Again, this also does not happen without a body, and I judge that I really did not sense those many things I seemed to have sensed in my dreams. How about thinking? Here I discover that thought is an attribute that really does belong to me. This alone cannot be detached from me. I am; I exist; this is certain. But for how long? For as long as I think. Because perhaps it could also come to pass that if I should cease from all thinking I would then utterly cease to exist. I now admit nothing that is not necessarily true. I am therefore precisely only a thing that thinks; that is, a mind, or soul, or intellect, or reason—words the meaning of which I was ignorant before. Now, I am a true thing, and truly existing; but what kind of thing? I have said it already: a thing that thinks.

What then? I will set my imagination going to see if I am not something more. I am not that connection of members which is called the human body. Neither am I some subtle air infused into these members, not a wind, not a fire, not a vapor, not a breath—nothing that I imagine to myself, for I have supposed all these to be nothing. The assertion stands: the fact still remains that I am something. But perhaps is it the case that, nevertheless, these very things which I take to be nothing (because I am ignorant of them) in reality do not differ from that self which I know? This I do not know. I shall not quarrel about it right now; I can make a judgment only regarding things which are known to me. I know that I exist; I ask now who is this "I" whom I know. Most certainly the knowledge of this matter, thus precisely understood, does not depend upon things that I do not yet know to exist. Therefore, it is not dependent upon any of those things that I feign in my imagination. But this word "feign" warns me of my error. For I would be feigning if I should "imagine" that I am something, because imagining is merely the contemplation of the shape or image of a corporeal thing. But I know now with certainty that I am, and at the same time it could happen that all these images—and, generally, everything that pertains to the nature of the body—are nothing but dreams. When these things are taken into account, I would speak no less foolishly were I to say: "I will imagine so that I might recognize more distinctly who I am," than were I to say: "Now I surely am awake, and I see something true, but because I do not yet see it with sufficient evidence, I will take the trouble of going to sleep so that my dreams might show this to me more truly and more evidently." Thus I know that none of what I can comprehend by means of the imagination pertains to this understanding that I have of myself. Moreover, I know that I must be most diligent about withdrawing my mind from these things so that it can perceive its nature as distinctly as possible.

28

But what then am I? A thing that thinks. What is that? A thing that doubts, understands, affirms, denies, wills, refuses, and which also imagines and senses.

It is truly no small matter if all of these things pertain to me. But why should they not pertain to me? Is it not I who now doubt almost everything, I who nevertheless understand something, I who affirm that this one thing is true, I who deny other things, I who desire to know more things, I who wish not to be deceived, I who imagine many things against my will, I who take note of many things as if coming from the senses? Is there anything in all of this which is not just as true as it is that I am, even if I am always dreaming or even if the one who created me tries as hard as possible to delude me? Are any of these attributes distinct from my thought? What can be said to be separate from myself? For it is so obvious that it is I who doubt, I who understand, I who will, that there is nothing through which it could be more evidently explicated. But indeed I am also the same one who imagines; for, although perhaps as I supposed before, no imagined thing would be wholly true, the very power of imagining does really exist, and constitutes a part of my thought. Finally, I am the same one who senses or who takes note of bodily things as if through the senses. For example, I now see a light, I hear a noise, I feel heat. These are false, since I am asleep. But I certainly seem to see, hear, and feel. This cannot be false: properly speaking, this is what is called "sensing" in me. But this is, to speak precisely, nothing other than thinking.

From these considerations I begin to know a little better who I am. But it still seems that I cannot hold back from believing that bodily things—whose images are formed by thought, and which the senses themselves examine—are much more distinctly known than this unknown aspect of myself which does not come under the imagination. And yet it would be quite strange if the very things which I consider to be doubtful, unknown, and foreign to me are comprehended by me more distinctly than what is true, what is known—than, in fine, myself. But I see what is happening: my mind loves to wander and does not allow itself to be restricted to the confines of truth. Let it be that way then: let us allow it the freest rein in every respect, so that, when we pull in the reins at the right time a little later, the mind may suffer itself to be ruled more easily.

Let us consider those things which are commonly believed to be the most distinctly comprehended of all: namely the bodies which we touch and see. But not bodies in general, for these generic perceptions are often somewhat more confused; rather let us consider one body in particular. Let us take, for instance, this piece of wax. It has very recently been taken from the honeycombs; it has not as yet lost all the flavor of its honey. It retains some of the smell of the flowers from which it was collected. Its color, shape, and size are obvious. It is hard and cold. It can easily be touched, and if you rap on it with a knuckle it makes a sound. In short, everything is present in it that appears to be needed in order that a body can be known as distinctly as possible. But notice that while I am speaking, it is brought close to the fire; the remaining traces of the honey flavor are purged; the odor vanishes; the color is changed;

the original shape disappears. Its magnitude increases, it becomes liquid and hot, and can hardly be touched; and now, when you knock on it, it does not emit any sound. Up to this point, does the same wax remain? One must confess that it does: no one denies it; no one thinks otherwise. What was there then in the wax that was so distinctly comprehended? Certainly none of the things that I reached by means of the senses. For whatever came under taste or smell or sight or touch or hearing by now has changed, yet the wax remains.

Perhaps the wax was what I now think it is: namely that it really never was the sweetness of the honey or the fragrance of the flowers, not this whiteness, not a figure, not a sound, but a body which a little earlier manifested itself to me in these ways, and now does so in other ways. But just what precisely is this thing which I imagine thus? Let us direct our attention to this and see what remains after we have removed everything which does not belong to the wax: only that it is something extended, flexible, and subject to change. What is this flexible and mutable thing? Is it not the fact that I imagine that this wax can change from a round to a square shape, or from the latter to a rectangular shape? Not at all: for I comprehend that the wax is capable of innumerable changes, yet I cannot survey these innumerable changes by imagining them. Therefore this comprehension is not accomplished by the faculty of imagination. What is this extended thing? Is this thing's extension also unknown? For it becomes larger in wax that is beginning to liquify, greater in boiling wax, and greater still as the heat is increased. And I would not judge rightly what the wax is if I did not believe that this wax can take on even more varieties of extension than I could ever have grasped by the imagination. It remains then for me to concede that I in no way imagine what this wax is, but perceive it by the mind only. I am speaking about this piece of wax in particular, for it is clearer in the case of wax in general. But what is this wax which is perceived only by the mind? It is the same that I see, touch, and imagine; in short it is the same as I took it to be from the very beginning. But we must take note of the fact that the perception of the wax is neither by sight, nor touch, nor imagination, nor was it ever so (although it seemed so before), but rather an inspection on the part of the mind alone. This inspection can be imperfect and confused, as it was before, or clear and distinct, as it is now, according to whether I pay greater or less attention to those things of which the wax consists.

But meanwhile I marvel at how prone my mind is to errors; for although I am considering these things within myself silently and without words, nevertheless I latch onto words themselves and I am very nearly deceived by the ways in which people speak. For we say that we see the wax itself, if it is present, and not that we judge it to be present from its

31

32

color or shape. Whence I might conclude at once: the wax is therefore known by eyesight, and not by an inspection on the part of the mind alone, unless I perhaps now might have looked out the window at the men crossing the street whom I say I am no less wont to see than the wax. But what do I see over and above the hats and clothing? Could not robots be concealed under these things? But I judge them to be men; thus what I believed I had seen with my eyes, I actually comprehend with nothing but the faculty of judgment which is in my mind.

But a person who seeks to know more than the common crowd should be ashamed of himself if he has come upon doubts as a result of an encounter with the forms of speech devised by the common crowd. Let us then go forward, paying attention to the following question: did I perceive more perfectly and evidently what the wax was when I first saw it and believed I had known it by the external sense—or at least by the common sense, as they say, that is, the imaginative power—than I know it now, after having examined more diligently both what the wax is and how it is known. Surely it is absurd to doubt this matter. For what was there in the first perception that was distinct? What was there that any animal could not have seemed capable of possessing? But when I distinguish the wax from its external forms, as if having taken off its clothes, as it were, I look at the naked wax, even though at this point there can be an error in my judgment; nevertheless I could not perceive it without a human mind.

33   But what am I to say about this mind, or about myself? For as yet I admit nothing else to be in me over and above my mind. What, I say, am I who seem to perceive this wax so distinctly? Do I not know myself not only much more truly and with more certainty, but also much more distinctly and evidently? For if I judge that the wax exists from the fact that I see it, certainly it follows much more evidently that I myself exist, from the fact that I see the wax. For it could happen that what I see is not truly wax. It could happen that I have no eyes with which to see anything. But it could not happen that, while I see or think I see (I do not now distinguish these two), I who think am not something. Likewise, if I judge that the wax exists from the fact that I touch it, the same thing will again follow: I exist. If from the fact that I imagine, or from whatever other cause, the same thing readily follows. But what I noted regarding the wax applies to all the other things that are external to me. Furthermore, if the perception of the wax seemed more distinct after it became known to me not only from sight or touch, but from many causes, how much more distinctly I must be known to myself; for there are no considerations that can aid in the perception of the wax or any other body without these considerations demonstrating even better the nature of my mind. But there are still so many other things in my

mind from which one can draw a more distinct knowledge of the mind, so that those things which emanate from a body seem hardly worth enumerating.

But lo and behold, I have arrived on my own at the place I wanted. 34 Since I know that bodies are not, properly speaking, perceived by the senses or by the faculty of imagination, but only by the intellect, and since, moreover, I know that they are not perceived by being touched or seen, but only insofar as they are expressly understood, nothing can be more easily and more evidently perceived by me than my mind. But because an established habit of belief cannot be put aside so quickly, it is appropriate to stop here, so that by the length of my meditation this new knowledge may be more deeply impressed on my memory.

Meditation Three: Concerning God, That He Exists

Now I will shut my eyes, I will stop up my ears, I will divert all my senses, I will even blot out from my thoughts all images of corporeal things—or at least, since the latter can hardly be done, I will regard these images as nothing, empty and false as indeed they are. And as I converse only with myself and look more deeply into myself, I will attempt to render myself gradually better known and familiar to myself. I am a thing that thinks, that is to say, a thing that doubts, affirms, denies, knows a few things, is ignorant of many things, wills, rejects, and also imagines and senses. As I observed earlier, although these things that I sense or imagine may perhaps be nothing at all outside me, nevertheless I am certain that these modes of thinking—which I call sensations and imaginations—insofar as they are only modes of thinking, are within me. 35

In these few words I have summed up what I truly know, or at least what, so far, I have observed that I know. Now I will ponder more carefully to see whether perhaps there are other things pertaining to me that, up to this time, I have not yet noticed. I am certain that I am a thing that thinks. But do I not therefore also know what is required so that I may be certain of something? Surely in this first instance of knowing, there is nothing else than a certain clear and distinct perception of what I affirm. Yet this would hardly be sufficient to render me certain of the truth of a thing, if it could ever happen that something that I perceive so clearly and distinctly were false. And thus I now seem to be able to posit as a general rule that what I very clearly and distinctly perceive is true.

Nevertheless, I have previously admitted many things as wholly certain and evident that nevertheless I discovered afterward to be doubtful. What sort of things were these? Why, indeed, the earth, the heavens, the stars, and all the other things that I perceived by means of my senses. But what did I perceive clearly regarding these things? Certainly the ideas or thoughts of these things hovered before my mind. But I do not now deny that these ideas are in me. However, there was something else that I affirmed and that, because of my habit of believing it, I thought was something I clearly perceived, but that, in fact, I did not perceive— namely, that certain things are outside me, things from which those ideas proceed and things to which they were entirely similar. But on this point I was mistaken; or, if I judged truly, this judgment did not result from the force of my perception.

36   What then? When I considered something very simple and easy in the areas of arithmetic or geometry, for example that two plus three together make five, and the like, did I not intuit them at least clearly enough that I might affirm them to be true? To be sure, I did decide later on that I must doubt these things for no other reason than that it entered my mind that some God could have given me a nature such that I might be deceived—even about matters that seemed most evident. For every time this preconceived opinion about the supreme power of God occurs to me, I am constrained to admit that, if he wishes, it is very easy for him to cause me to err, even in those matters that I think I have in- tuited as plainly as possible with the eyes of the mind. Yet every time I turn my attention to those very things that I think I perceive with such great clarity, I am so entirely persuaded by these things that I sponta- neously burst out with these words: "let him who can deceive me; as long as I think that I am something, he will never bring it about that I am nothing, or one day make it true that I never existed, because it is true now that I am; nor will he ever bring it about that two plus three yield more or less than five, or that similar matters, in which I recognize an obvious contradiction, exist." And, certainly, because I have no occasion for thinking that there is a God who deceives, and because I am not cer- tain whether there even is a God, the reason for doubting—depending as it does on the above hypothesis—is very tenuous and, so to speak, meta- physical. Moreover, in order to remove this doubt, I ought at the first opportunity to inquire if there is a God, and, if there is, whether or not he can be a deceiver. If I am ignorant of these matters, I do not think I can ever be certain of anything else.

37   Now, however, good order seems to demand that I first arrange all my thoughts into certain classifications, and inquire in which of these truth or falsity properly consists. Some of these thoughts are like images of things. To these alone does the word "idea" properly apply, for exam- ple, when I think of a man, a chimera, heaven, an angel, or God. Again,

there are other thoughts that take different forms: when I will, when I fear, when I affirm, when I deny, there is always something I grasp as the subject of my thought, yet I comprehend something that is more than the mere likeness of a thing. Some of these thoughts are called volitions or emotions, while others are called judgments.

Now as to what pertains to ideas, if they are considered alone and in their own right, without being referred to something else, they cannot then properly be considered false. For whether I imagine a she-goat or a chimera, it is no less true that I imagine the one than the other. Moreover, we need not fear that falsity exists in the will itself or in the emotions; although I can choose evil things or even things that nowhere exist, it does not follow that it is untrue that I do choose these things. Thus there remain only judgments. I must take care not to be mistaken about these. Now the principal and most frequent error that can be found in judgments consists in the fact that I judge that the ideas, which are in me, are similar to, or in conformity with, certain things outside me. Indeed, if I consider these ideas only as certain modes of my thought, and do not refer them to something else, they can hardly give me any cause for error.

Among these ideas, some seem to me to be innate, some seem to be derived from an external source, and some seem to be produced by me. I understand what a thing is, what truth is, what thought is—I do not seem to have derived these from any source other than from my very own nature. But now I hear a noise, I see the sun, and I feel a fire; until this point I judged that these things proceeded from certain things existing outside me. Finally, I judged that sirens, hippogriffs, and the like have been formed by me. Or perhaps I could also suppose that all of these ideas are derived from an external source, or are innate, or are produced, for I have not yet ascertained clearly their true origin.

But in this meditation those ideas that I believe to be drawn from things existing outside me must especially be investigated. Just what reason is there that moves me so that I believe that these ideas are similar to those objects? Well, I seem to be taught that way by nature. Moreover, I know by experience that these ideas do not depend upon my will, nor consequently upon myself, for often I notice them against my will; as now—whether I will it or not—I feel heat, and therefore I believe that this feeling or idea of heat comes to me from something other than myself, namely from the fire I am near. Nothing is more obvious than the judgment that this object (rather than something else) grafts its likeness on to me.

I will now see whether these reasons are valid enough. When I say in this meditation that I have been taught so by nature, I understand only that I am driven by a spontaneous impulse to believing this position, and

38

not that some light of nature shows me it is true. These two positions are at considerable odds with one another. For whatever this light of nature shows me—for example, that from the fact that I doubt, it follows that I am and so on—cannot in any way be doubtful, because there can be no other faculty in which I may trust as much as the light of nature that could teach which of these positions are not true. As for natural impulses, I have very often judged myself to have been driven by them to the poorer choice, when it was a question of choosing a good; I do not see why I should trust them any more in other matters.

Again, although these ideas do not depend upon my will, it does not therefore follow that they necessarily proceed from things existing outside me. Although these impulses I have discussed are in me, nevertheless they seem to be different from my will. Thus perhaps there is also in me some other faculty, one not yet sufficiently known to me, that produces these ideas, because it has always seemed that these ideas are formed in me when I sleep, without any aid from external things.

Again, even if these ideas do proceed from things other than myself, it does not therefore follow that they must be similar to these objects. Indeed it seems I have frequently noticed a vast difference in many respects. For example, I find within myself two distinct ideas of the sun. One idea derives, as it were, from the senses. Now it is this idea that, among those things I consider to be derived from outside me, is most in need of examination. By means of this idea the sun appears to me to be very small. Yet there is another idea, one derived from the computations of astronomy. This idea is elicited from certain notions innate in me (or is somehow made by me) through which the sun is shown to be several times larger than the earth. Both ideas surely cannot be similar to the same sun existing outside me, and reason convinces me that the one that seems to have emanated from it at such close proximity is the very one that is most dissimilar to the sun.

All of this demonstrates sufficiently that up to this point I have believed not by certain judgment, but only by a blind impulse that things exist outside me that send their ideas or images into me through the sense organs or by some other means.

But still another path occurs to me for inquiring whether there are some objects, from among those things of which there are ideas in me, that exist outside me. Now, insofar as these ideas are merely modes of thought, I do not see any inequality among them; they all seem to proceed from me in the same way. But insofar as one idea represents one thing and another idea another thing, it is obvious that they are very different from one another. There is no question that those ideas that exhibit substances to me are something more and, as I phrase it, contain more objective reality[1] in themselves than those which represent only modes or accidents. Again, the idea that enables me to understand a highest God, one who is eternal, infinite, omniscient, omnipotent, and creator of all things other than himself, has more objective reality in it than those ideas through which finite substances are exhibited.

But it is evident by the light of nature that at the very least there must be as much in the total efficient cause as there is in the effect of that same cause. For, I ask, where can an effect get its reality unless it be from its cause? And how can the cause give that reality to the effect, unless the cause also has that reality? Hence it follows that something cannot come into existence from nothing, nor even can what is more perfect, that is, that contains in itself more reality, come into existence from what contains less. But this is clearly true not merely for those effects whose reality is actual or formal, but also for ideas in which only objective reality is considered. This means, for example, that a stone, which did not exist before, can in no way begin to exist now, unless it be produced by something in which there is, either formally or eminently,[2] everything that is in the stone; nor can heat be introduced into a subject which was not hot before unless it is done by something that is of at least as perfect an order as heat—the same holds true for the rest. Moreover, in me there can be no idea of heat, or of a stone, unless it comes to me from some cause that has at least as much reality as I conceive to be in the heat or in the stone. Although this cause transmits none of its actual or formal reality into my idea, it is not to be thought that the idea must be less real. Rather, the very nature of this idea is such that it needs no formal reality other than itself—except what it borrows from my thought, of which it is a mode. That this idea contains this or that objective reality rather than some other one results from the fact that the idea gets its objective reality from a cause in which there is at least as much formal

41

---

[1] By "objective reality" Descartes means the content of an idea, or an object insofar as it is represented in an idea. By "formal reality" he means the object insofar as it actually exists in its own right, that is, independently of its being perceived.

[2] By a "formal cause" Descartes means a cause that contains exactly the same degree or amount of reality and perfection as its effect. For example, a human being in generating another human being is said to be the "formal cause" of the generated human being. Again, were an object, such as a tree stump, to cause in me a perception of a tree stump, then the object is said to be the "formal cause" of my perception. By an "eminent cause" Descartes means a cause that contains more reality and perfection than its effect. For example, God, in creating a human being, would be said to be the "eminent cause" of that human being because there is clearly a greater degree or amount of reality and perfection in God than in a human being. Again, were God to cause in me a perception of a tree stump, then God would be said to be the "eminent cause" of my perception of the tree stump.

reality as there is objective reality contained in the idea. For if we posit that something is found in the idea that was not in its cause, then the idea would get it from nothing; but as imperfect a mode of being as this is, by which a thing exists in the intellect objectively through an idea, it nevertheless is surely not nothing; hence it cannot get its existence from nothing.

Also, because the reality that I consider in my ideas is only objective, I should not suspect that there is no need for the same reality to be formally in the causes of these ideas, but that it suffices for it to be in them objectively. To the extent that the objective mode of being pertains to ideas by their very nature, so the formal mode of being belongs to the causes of ideas, at least to the first and preeminent ones, by their very nature. Although one idea can perhaps come into being from another, nevertheless there is no infinite regress here; at length some first idea must be reached whose cause is a sort of archetype that formally contains all the reality that is in the idea only objectively. Thus it is evident to me by the light of nature that my ideas are like images that can easily fail to match the perfection of the things from which they have been drawn, but my ideas cannot contain anything better or more perfect.

But the longer and more attentively I examine all these points, the more clearly and distinctly I know they are true. But what do I finally conclude? Why, if the objective reality of one of my ideas is such that I am certain that the same reality is not formally or eminently in me, and that therefore I myself cannot be the cause of the idea, then it necessarily follows that I am not alone in the world, and that something else—the cause of this idea—also exists. If in fact no such idea is found in me, I shall plainly have no argument to make me certain of the existence of something other than myself. For I have looked at all of these arguments most diligently and so far I have been unable to find any other.

Among my ideas, in addition to the one that represents me to myself—about which there can be no difficulty at this point—there are some that represent God, others that represent inanimate, corporeal objects, others that represent angels, others that represent animals, and finally others that represent men like myself.

As to the ideas that represent other men, animals, or angels, I easily understand that they can be formed from the ideas that I have of myself, of corporeal things, and of God—even if no men except myself, no animals, and no angels were to exist in the world.

As to the ideas of corporeal things, there is nothing in them which is such that it seems unable to have come from me. For if I investigate thoroughly and if I examine each one individually in the same way that I examined the idea of wax yesterday, I notice that there are only a very few properties that I perceive in them clearly and distinctly: namely, magnitude, or extension in length, breadth, and depth; shape, which arises from the limit of this extension; position, which the various shaped things possess in relation to one another; and motion, or the alteration of this position; to these can be added substance, duration, and number. The remaining properties, however—such as light and colors, sounds, odors, tastes, heat and cold and other tactile qualities—are thought by me only in a very confused and obscure manner, with the result that I do not know whether they are true or false, that is, whether the ideas that I have of these things are ideas of certain things or are not ideas of things. For although a short time ago I noted that falsity properly so called ("formal" falsity) can be found only in judgments, nevertheless there is another kind of falsity (called "material" falsity) in ideas; it occurs whenever judgments present a non-thing as if it were a thing. For example, the ideas I have of heat and cold fall so short of being clear and distinct that I cannot learn from them whether cold is only a privation of heat or whether heat is a privation of cold; whether both are real qualities or neither is. Because ideas can only be ideas of things, and if it is true that cold is nothing more than the privation of heat, then an idea such as this one, that represents something real and positive to me, will not inappropriately be called false—and so too for the other ideas.

Assuredly it is not necessary for me to assign to these ideas an author distinct from me. For if they were false, that is, if they should represent no objects, I know by the light of nature that they proceed from nothing; that is, they are in me for no other reason than that something is lacking in my nature, that my nature is plainly not perfect. If, on the other hand, these ideas are true because they exhibit so little reality to me that I cannot distinguish them from a non-thing, then I see no reason why they cannot get their existence from myself.

As for those things that are the clear and distinct elements in the ideas of corporeal things, I seem able to have borrowed some from the idea of myself: namely, substance, duration, number, and whatever else there may be of this type. For when I think that a stone is a substance, that is to say, a thing that in its own right has an aptitude for existing, and that I too am a substance—although I conceive that I am a thing that thinks and not an extended thing, whereas a stone is an extended thing and not a thing that thinks—there is, accordingly, the greatest diversity between these two concepts, even though they seem to agree with one another when they are considered under the rubric of substance. Furthermore, when I perceive that I exist now and recall that I have previously existed for some time, and when I have several thoughts and know the number of these thoughts, I acquire the ideas of duration and number, which I then apply to everything else. However, all the other elements out of which the ideas of corporeal things are put together (namely extension, shape, position, and motion) are not contained in me formally, because I am only a thing that thinks. But because they are only modes of a substance, and I too am a substance, they seem capable of being contained in me eminently.

Thus there remains only the idea of God. We must consider whether there is in this idea something which could not have originated from me. I understand by the word "God" an infinite and independent substance, intelligent and powerful in the highest degree, who created me along with everything else—if in fact there is anything else. Indeed all these qualities are such that, the more diligently I attend to them, the less they seem capable of having arisen from myself alone. Thus, from what has been said above, we must conclude that God necessarily exists.

For although the idea of substance is in me by virtue of the fact that I am a substance, nevertheless it would not for that reason be the idea of an infinite substance, unless it proceeded from some substance which is in fact infinite, because I am finite.

Nor should I think that I do not perceive the infinite by means of a true idea, but only through a negation of the finite, just as I perceive rest and shadows by means of a negation of motion and light. On the contrary, I clearly understand that there is more reality in an infinite substance than there is in a finite one. Thus the perception of the infinite somehow exists in me prior to the perception of the finite, that is, the perception of God exists prior to the perception of myself. Why would I know that I doubt and I desire, that is, that I lack something and that I am not wholly perfect, if there were no idea in me of a more perfect being by comparison with which I might acknowledge my defects?

Nor can it be said that this idea of God is perhaps materially false, and therefore can be from nothing, as I pointed out just now regarding the ideas of heat and cold and the like. On the contrary, because it is the most clear and distinct of all ideas and because it contains more objective reality than any other idea, no idea is truer in its own right, and there is no idea in which less suspicion of falsity is to be found. I maintain that this idea of a being most perfect and infinite is true in the highest degree. Although such a being can perhaps be imagined not to exist, it nevertheless cannot be imagined that this idea shows me nothing real, as was the case with the idea of cold that I referred to earlier. It is also an idea that is clear and distinct in the highest degree; for whatever I clearly and distinctly perceive that is real and true and that contains some perfection is wholly contained in that idea. It is not inconsistent to say that I do not comprehend the infinite or that there are countless other things in God that I can in no way either comprehend or perhaps even touch with thought. For the nature of the infinite is such that it is not comprehended by me, who am finite. And it is sufficient that I understand this very point and judge that all those things that I clearly perceive and that I know to contain some perfection—and perhaps even countless other things of which I am ignorant—are in God either formally or eminently.

The result is that, of all those that are in me, the idea that I have of him is the most true, the most clear, and the most distinct.

But perhaps I am something greater than I take myself to be. Perhaps all these perfections that I attribute to God are somehow in me potentially, although they do not yet assert themselves and are not yet reduced to act. For I now observe that my knowledge is gradually being increased; I see nothing that stands in the way of my knowledge being increased more and more to infinity. I see no reason why, with my knowledge thus increased, I cannot acquire all the remaining perfections of God. And, finally, if the potential for producing these perfections is in me already, I see no reason why this potential does not suffice to produce the idea of these perfections.

Yet none of these things can be the case. First, while it is true that my knowledge is gradually increased and that in me there are many elements in potency which do not yet exist in act, nevertheless, none of these elements pertain to the idea of God, in which nothing whatever is potential; this gradual increase is itself a most certain argument for my imperfection. Moreover, although my knowledge might always increase more and more, nevertheless I understand that this knowledge will never by this means be infinite in act, because it will never reach a point where it is incapable of greater increase. On the contrary, I judge God to be infinite in act, with the result that nothing can be added to his perfection. Finally, I perceive that the objective being of an idea cannot be produced by a merely potential being (which, properly speaking, is nothing), but only by an actual or formal being.

Indeed there is nothing in all these matters that is not manifest by the light of nature to a person who is diligent and attentive. But when I am less attentive, and the images of sensible things blind my powers of discernment, I do not so easily recall why the idea of a being more perfect than me necessarily proceeds from a being that really is more perfect. This being the case, it is appropriate to ask further whether I myself who have this idea could exist, if such a being did not exist.

From what source, then, do I derive my existence? From myself, from my parents, or from whatever other things there are that are less perfect than God. Nothing more perfect than He, nothing even as perfect as He, can be thought or imagined.

But if I were derived from myself, I would not doubt, I would not hope, and I would not lack anything whatever. For I would have given myself all the perfections of which I have some idea; in so doing, I would be God! I must not believe that these things that I lack are perhaps more difficult to acquire than those which I now have. On the contrary, it is evident that it would have been much more difficult for me—that is, a thing or substance that thinks—to emerge from nothing than it would be to acquire the knowledge—something that is only an accident of this substance—of the many things about which I am ignorant. Certainly, if I received this greater quality from myself, I would not have denied myself at least those things that can be had more easily. Nor would I scarcely deny myself any of those other things that I perceive to be contained in the idea of God, because surely nothing seems to me more difficult to make. But if there were any of those other things that were more difficult to make, then they certainly would also seem more difficult to me, assuming, of course, that the remaining ones that I possess I have from myself, because I observe that with them my power is terminated.

Nor do I avoid the cogency of these arguments, if I suppose that perhaps I have always been as I am now, as if it then followed that no author of my existence need be sought. Because the entire period of one's life can be divided into countless parts, each of which in no way depends on the others, it does not follow from the fact that I existed a short while ago that I now ought to exist, unless some cause creates me once again, as it were, at this moment—that is to say, preserves me. For it is obvious to one who is cognizant of the nature of time that the same force and action is needed to preserve anything at all during the individual moments that it lasts as is needed to create that same thing anew—if it should happen not yet to exist. It is one of those things that is manifest by the light of nature that preservation differs from creation solely by virtue of a distinction of reason.

Therefore I ought now to ask myself whether I have some power through which I can bring it about that I myself, who now am, will also exist a little later? Because I am nothing but a thing that thinks—or at least because I am now dealing only with precisely that part of me that is a thing that thinks—if such a power were in me, then I would certainly be aware of it. But I observe that there is no such power; from this fact I know most evidently that I depend upon a being other than myself.

But perhaps this being is not God, and I have been produced either by my parents or by some other causes less perfect than God. Indeed, as I said before, it is evident that there ought to be at least as much in the cause as there is in the effect. Thus, because I am a thing that thinks and because I have within me an idea of God—whatever cause is at length assigned to me—it must be granted that it too is a thing that thinks and has an idea of all the perfections which I attribute to God. And one can again inquire whether it exists through itself or through another. For if it derived from itself, it is evident from what has been said that this thing is God, because, having the power of existing in its own right, it also unquestionably has the power of possessing in act all the perfections whose idea it has in itself—that is, all of those perfections that I conceive as being in God. However, if it is from another cause, one will then once again inquire in a similar fashion about this other thing—whether it is through itself or through another—until finally one arrives at the ultimate cause, which will be God.

For it is apparent enough that there can be no infinite regress, especially since I am dealing here not only with the cause that once produced me, but also and most especially with the cause that preserves me at the present time.

Nor can one imagine that perhaps several partial causes have concurred to bring me into being, and that from one cause I have taken the idea of one of the perfections I attribute to God and from another cause the idea of another perfection; so that all of these perfections are found somewhere in the universe, but not all joined together in one thing, which would be God. On the contrary, unity, simplicity, or the inseparability of all those things which are in God is one of the principal perfections that I understand to be in him. Certainly the idea of the unity of all these perfections could not have been placed in me by a cause from which I do not also have the ideas of the other perfections; for it could not bring it about that I would instantaneously understand them to be joined to one another inseparably, unless it brought it about instantaneously that I recognize which ones these are.

Finally, as to my parents, even if everything that I ever believed about them were true, still they do not preserve me; nor did they bring me into being, insofar as I am a thing that thinks. Rather, they merely placed certain dispositions in the matter in which I judged that I—that is, a mind that is all that I now accept for myself—am contained. And thus there can be no difficulty here concerning these matters. Rather, one has no choice but to conclude that, from the simple fact that I exist and that an idea of a most perfect being, that is, God, is in me, it is most evidently demonstrated that God exists.   **51**

There remains only the question of how I received this idea of God. For I did not draw it from the senses, and it never came to me when I did not anticipate it, as the ideas of sensible things are wont to do when these things present themselves—or seem to present themselves—to the organs of the senses. Nor has it been produced by me, for I plainly cannot add or subtract anything from it. It thus remains that it is innate in me, just as the idea of myself is also innate in me.

To be sure it is not astonishing that in creating me, God endowed me with this idea, so that it would be like the sign of an artist impressed upon his work. Nor is it necessary that this sign be something distinct from the work itself. But from this one fact that God created me, it is highly believable that I have somehow been made in his image and likeness, and that I perceive this likeness (in which the idea of God is contained) by means of the same faculty through which I perceive myself. That is, when I turn my powers of discernment toward myself, I not only understand that I am something incomplete and dependent upon another, something aspiring indefinitely for greater and greater or better things, but instantaneously I also understand that the thing on which I depend has in himself all of these greater things—not merely indefinitely and potentially, but infinitely and actually. Thus it is God. The whole force of the argument rests on the fact that I recognize that it is impossible that I should exist, being of such a nature as I am—namely, having the idea of God in me—unless God in fact does exist. God, I say, that same being whose idea is in me: a being having all those perfections that I cannot comprehend, but in some way can touch with my thought, and a being subject to no defects. From these things it is sufficiently obvious that he cannot be a deceiver. For it is manifest by the light of nature that fraud and deception depend on some defect.

But before examining this idea more diligently and at the same time inquiring into the other truths that can be inferred from it, it is appropriate at this point to spend some time contemplating this God, to consider within myself his attributes and the beauty of this immense light, so far as the power of discernment in my darkened wit can carry me, to gaze, to admire, and to adore. For just as we believe by faith that the greatest felicity of the next life consists in nothing more than this contemplation of the divine majesty, so now, from the same—though much less perfect—contemplation we observe that the greatest pleasure of which we are capable in this life can be perceived.

Meditation Four: Concerning the True and the False

Thus have I lately become accustomed to withdrawing the mind from the senses. I have so carefully taken note of the fact that there are very few things that are truly perceived regarding corporeal things, and that a great many more things regarding the human mind—and yet still many more things regarding God—can be known, that with no difficulty do I direct my thinking away from things I can imagine to things that I can only understand, and that are separated from all matter. And I have a manifestly much more distinct idea of the human mind, insofar as it is a thinking thing—not extended in length, breadth or depth—receiving nothing else from a body except the idea of this or that body. And when I direct my attention to the fact that I am doubting, that is, that I am an incomplete and dependent thing, there comes to mind a clear and distinct idea of an independent and complete being, namely, God. And from the mere fact that this idea is in me, or that I who have this idea exist,

I plainly conclude that God also exists, and that my existence depends entirely upon him at each and every moment. The result is that I am confident that nothing more evident, nothing more certain can be known by the human mind. From this contemplation of the true God, in whom are hidden all the treasures of knowledge and wisdom, I now seem to see a way by which I might attain knowledge of other matters.

To begin with, I acknowledge that it is impossible for him ever to deceive me; for in every trick or deception some imperfection is found. Although the ability to deceive seems to be an indication of cleverness or power, nevertheless, willful deception evinces maliciousness and weakness. Accordingly, deception is not compatible with God.

Next I observe that there is in me a certain faculty of judgment that I undoubtedly received from God, as is the case with all the other things that are in me. Since he has not wished to deceive me, he certainly has not given me a faculty such that, when I use it properly, I could ever make a mistake.

No doubt regarding this matter remains, except that then it seems to follow that I should never make a mistake, for if all that I have is from God—and he gave me no faculty of making a mistake—I seem incapable of ever erring. So, as long as I think about God alone and direct myself straightforwardly and directly at him, I discover no grounds for error or falsity. But, after I turn my attention back on myself, I nevertheless observe that I am subject to countless errors. Seeking a cause for these errors, I notice that not only a real and positive idea of God (that is, of a being supremely perfect), but also, so to speak, a certain negative idea of nothing (that is, of what is supremely lacking in every perfection) passes before me. I also notice that I have been so constituted as to be some kind of middle ground between God and nothing, or between supreme being and non-being, so that, insofar as I have been created by a supreme being, there is nothing in me by means of which I might be deceived or be led into error; but insofar as I am not the supreme being, I lack quite a few things—so much so that it is not surprising that I am deceived. Thus I certainly understand that error as such is not something real that depends upon God, but rather is only a defect. Nor for that reason is there a need to account for my errors by positing a faculty given to me by God. Rather, it happens that I make mistakes because the faculty of judging the truth, which I have from him, is not infinite in me.

Still, this is not yet altogether satisfactory; for error is not a pure negation, but a privation or a lack of some knowledge that somehow ought to be in me. It does not seem possible to one who attends to the nature of God that he could have placed in me a faculty that is not perfect within its class or that has been deprived of some perfection it ought to have. For if it is true that the more perfect the artisan, the more perfect

54

55

the works he produces, then what can be made by that supreme creator of all things that is not perfect in all respects? No doubt God could have made me such that I never err; no doubt, again, God always wills what is best; is it not therefore better for me to be deceived than not to be?

While I mull over these things more carefully, it immediately dawns on me that there is no reason to marvel at the fact that God might make things the reasons for which I do not understand; his existence ought not therefore to be doubted, because I might happen to observe that there are other things that were made by him whose "why" and "how" I do not comprehend. Since I know now that my nature is very weak and limited, but that the nature of God is immense, incomprehensible, and infinite, therefore I also know with sufficient evidence that he can make innumerable things whose causes escape me. For this reason alone the whole class of causes, which people customarily derive from a thing's "purpose," I judge to be useless in dealing with matters related to physics. It is not without rashness that I think myself competent to inquire into God's purposes.

It also occurs to me that whenever we ask whether the works of God are perfect, we should examine the whole universe together and not just one creature in isolation from the rest. Something, if all by itself, may rightfully appear very imperfect; but if it is seen in its role as a part in the universe, it is most perfect. However, from having wanted to doubt everything, I know with certainty only that I and God exist; nevertheless, from having taken note of the immense power of God, I am unable to deny that many other things have been made by him, or at least can be made by him, and, what is more, that I would take on the added significance of being a part in the universe of things.

Finally, focusing closer on myself and inquiring into the nature of my errors (the only things that argue for an imperfection in me), I note that these errors depend on the simultaneous concurrence of two causes: the faculty of knowing that is in me, and the faculty of choosing (in other words, the free choice of the will), that is, they depend on the intellect and will at the same time. Through the intellect alone I perceive only ideas concerning which I can make a judgment; no error, properly so-called, is to be found in the intellect thus precisely conceived. For although countless things about which I have no idea perhaps exist, nevertheless I must not be said, properly speaking, to be deprived of them. I am bereft of them only in a negative way, because I can give no argument by which to prove that God ought to have given me a greater faculty of knowing than he has. No matter how expert a craftsman I understand him to be, still I do not therefore believe he ought to have bestowed on each one of his works all the perfections that he can put into some. In fact, I cannot complain that I have received from God an in

sufficiently ample and imperfect will, or free choice, because I observe
that it is limited by no boundaries. And it seems eminently worth noting   57
that nothing else in me is so perfect or so great that I do not understand
how they can be even more perfect or greater. If, for example, I consider the faculty of understanding, I immediately recognize that it is very
small and quite finite in me, and at the same time I form an idea of another
much greater faculty—in fact, the greatest and infinite; and from the fact
that I can form an idea of this faculty, I perceive that it pertains to the
nature of God. Similarly, if I examine the faculty of memory, imagination, or any other faculty, I manifestly find none, except what in me is
feeble and limited, but what in God I understand to be immense. The
sole exception is the will or free choice; I observe it to be so great in me
that I grasp an idea of nothing greater, to the extent that the will is
principally the basis for my understanding that I bear an image and likeness of God. Although it is incomparably greater in God than it is in
me, both by virtue of the knowledge and power that are associated with
it and that render it more resolute and efficacious—as well as by virtue
of its object, for the divine will extends to more things; yet taken formally, will precisely as such, it does not seem greater, because willing is
only a matter of being able to do or not do something (that is, of being
able to affirm or deny, to pursue or to shun), or better still, the will consists solely in the fact that when something is proposed to us by our intellect either to affirm or deny, to pursue or to shun, we are moved by
it in such a way that we sense that no external force could have imposed it
on us. In order to be free it is unnecessary for me to be moved in [n]either
direction; on the contrary, the more I am inclined toward one direction—either because I evidently grasp that there is in it an aspect of the
good and the true, or because God has thus disposed the inner recesses   58
of my thought—the more I choose that direction more freely. Nor indeed
does divine grace or natural knowledge ever diminish one's liberty; rather
it increases and strengthens it. However, the indifference that I observe
when no reason moves me more in one direction than in another is the
lowest level of freedom; it evinces no perfection in it, but rather a defect
in my knowledge, or a certain negation. Were I always to see clearly
what is true and good, I would never deliberate about what is to be
judged or chosen. Thus, although I may be entirely free, I could never
for that reason be indifferent.

But from these considerations I perceive that the power of willing,
which is from God, is not—taken by itself—the cause of my errors, for in
its own way it is most ample and perfect. The same goes for the power
of understanding. Whatever I understand, because it is from God that I
have the power of understanding, I doubtless understand rightly; it is
impossible for anything to take place in the intellect that could cause me

to be deceived. From what source, therefore, do my errors arise? Solely from the fact that, because the will extends further than the intellect, I do not contain the will within the same boundaries; rather, I even extend it to things I do not understand. Because my will is indifferent to these latter things, it easily turns away from the true and the good; in this way I am deceived and commit sin.

For example, during these last few days when I have been examining whether anything exists in the world, I have noticed that, from the very fact that I had made this examination, it evidently followed that I exist. I could not help judging that what I understood clearly is true; not that I was coerced into holding this judgment because of some external force, but because a great inclination of my will followed from a great light in the intellect—so much so that the more spontaneously and freely I believed it, the less I was indifferent to it. But now I not only know that I, insofar as I am a thing that thinks, exist, but I also know that, having observed some ideas of corporeal nature, I might question whether the thinking nature that is in me—or rather that I am—is something different from this corporeal nature, or whether both natures are the same thing. And I presume that no consideration has as yet occurred to my mind which convinces me of the one more than the other. Thus it certainly follows that I am indifferent about affirming or denying either one, or even about my making any judgment at all in the matter.

Moreover, this indifference extends not only to those things about which the intellect knows absolutely nothing, but generally to everything of which the intellect does not have a keen enough knowledge at the very time when the will is deliberating on them. Although probable guesses might lead me in one direction, all it takes to move me to assent to the very opposite is the knowledge that they are merely guesswork, not certain and indubitable proofs. These last few days I have ample experience of this point, since everything that I had once believed to be as true as it possibly could be, I have now presumed to be utterly false, for the sole reason that I noticed that I could one way or another raise doubts about it.

But if I hold off from making a judgment when I do not perceive with sufficient clarity and distinctness what is in fact true, I clearly would be acting properly and would not be deceived. But were I to make an assertion or a denial, then I would not be using my freedom properly. If I turn in the direction that happens to be false, I am plainly deceived. But if I should embrace either alternative, and in so doing happen upon the truth by accident, I would still not be without fault, for it is manifest by the light of nature that the intellect's perception must always precede the will's being determined. Inherent in this incorrect use of the free will is a privation that constitutes the very essence of error: a privation, I say, is inherent in this operation insofar as the operation proceeds from me, but not in the faculty that was given to me by God, nor even in the operation of the faculty insofar as it depends upon God.

Indeed, I have no cause for complaint on the grounds that God has not given me a greater power of understanding or a greater light of nature than he has given me, for it is of the very essence of a finite intellect not to understand many things, and it is of the very essence of a created intellect to be finite. Actually, rather than think that he has withheld from me or deprived me of those things that he has not given, I ought to thank him, who never owed me anything, for what he has given me.

Again, I have no cause for complaint on the grounds that God has bestowed upon me a will that is open to more things than my intellect; since the will consists of merely one thing, and, as it were, something indivisible, it does not seem that its nature could withstand anything being removed from it. Indeed, the more ample the will is, the more I ought to thank its giver.

Nor again should I complain because God concurs with me in eliciting those acts of the will, or those judgments, in which I am deceived. For those acts are absolutely true and good, insofar as they depend on God; and it is, so to speak, a greater perfection in me in that I can elicit those acts than if I could not do so. But a privation, in which alone the meaning of falsehood and sin is to be found, in no way needs the concurrence of God, for a privation is not a thing; and it is not related to God as its cause; rather, it ought to be called only a negation. For it is no imperfection in God that he gave me the freedom to give or withhold my assent to things of which he has placed no clear and distinct perception in my intellect. But surely it is an imperfection in me that I not use my liberty well and that I make a judgment about what I do not rightly understand. Nevertheless, I see that God could easily have brought it about that, while remaining free and having finite knowledge, I still never err. This result could occur either had he given my intellect a clear and distinct perception of everything about which I would ever deliberate or had he only impressed firmly enough upon my memory—so that I could never forget it—that I should never judge anything that I do not clearly and distinctly understand. I readily understand that, insofar as I am something whole and complete in my own right, I would have been more perfect than I now am, were God to have made me that way. But I cannot therefore deny that it might be, so to speak, a greater perfection in the universe as a whole that some of its parts are not immune to error, while others are, than if they were all alike. And I have no right to complain that God wished to sustain me in the world as a person who is not the principal and most perfect of all.

Furthermore, even if I cannot abstain from errors in the first way, which depends upon the evident perception of everything concerning which one must deliberate, nevertheless I can do so by the second way, which depends solely on my remembering that whenever the truth of a given matter is not apparent, I must abstain from making judgments. Although I observe that there is in me this infirmity, namely that I am unable always to adhere fixedly to one and the same knowledge, nevertheless I can, by attentive and frequently repeated meditation, bring it to pass that I recall it every time the situation demands; thus, I would acquire a habit of not erring.

For this reason, since the greatest and chief perfection of man consists in investigating the cause of error and falsity, I think that today's meditation was quite profitable. Nor can this perfection be anything other than I have described; for whenever I restrain my will in making judgments, so that it extends only to those matters that are clearly and distinctly shown to it by the intellect, it can never happen that I err, because every clear and distinct perception is surely something. Hence it cannot derive its existence from nothing; rather, it necessarily has God for its author—that supremely perfect God, I say, to whom it is repugnant to be a deceiver—thus the perception is surely true. Nor today have I learned only what I must avoid so as not to be deceived; I have also simultaneously learned what I must do to search for the truth. I search for it indeed if only I attend enough to what I perfectly understand, and refrain from the rest that I apprehend more confusedly and more obscurely. I shall give this matter careful attention later.

63   Meditation Five: Concerning the Essence of Material Things, and Again Concerning God, That He Exists

Many matters still remain for me to examine concerning the attributes of God and of me, that is, concerning the nature of my mind. But perhaps I might take them up at some other time. For now, nothing seems more urgent—having noted what is to be avoided and what is to be done in order to attain the truth—than that I try to free myself from the doubts into which I have recently fallen, and that I see whether or not anything certain can be obtained concerning material things.

Before inquiring whether any such material things exist outside me, surely I ought to consider the ideas of those things, insofar as they are in my thought, and t see which ones are distinct and which ones are confused.

To be sure, I distinctly imagine that quantity, which philosophers commonly call "continuous": namely, the extension of its quantity, or rather the extension of the thing having quantitative dimensions of length, breadth, and depth. I enumerate the thing's various parts. I ascribe to these parts certain sizes, shapes, positions, and movements from place to place; to these movements I ascribe various durations.

Not only have I plainly known and examined those things, viewed thus in a general way, but also I perceive attentively the countless particular aspects of figures, number, movement, and so on. Their truth is open and suitable to my nature to such an extent that, when I first discover them, it is not so much that I seem to learn anything new but that I recall something I already knew. That is, I first notice things that were already in me, although I had not directed a mental gaze toward them.

This is what I believe most needs examination here: I find within me countless ideas of things, that, although perhaps not existing anywhere outside me, still cannot be said to be nothing. Although I somehow think them at will, nevertheless I have not put them together; rather, they have their own true and immutable natures. For example, when I imagine a triangle, although perhaps no such figure exists outside my thought anywhere in the world and never will, still its nature, essence, or form is completely determined, unchangeable, and eternal. I did not produce it and it does not depend on my mind. This point is evident from the fact that many properties can be demonstrated regarding this triangle—namely that its three angles are equal to two right angles, that its longest side is opposite the largest angle, and so on; whether I want to or not, I now clearly acknowledge them, although I had not previously thought of them at all when I imagined a triangle. For this reason, then, I have not produced them.

It is irrelevant to say that perhaps the idea of a triangle came to me from external things through the sense organs, on the grounds that on occasion I have seen triangular-shaped bodies. For I can think of many other figures, concerning which there can be no suspicion of their ever having entered me through the senses, and yet the various properties of these figures, no less than those of the triangle, can be demonstrated. All of these are patently true because they are clearly known by me; thus they are something and not merely nothing. For it is evident that everything that is true is something; I have already demonstrated at length that all that I know clearly is true. Had I not demonstrated this, certainly the nature of my mind is such that I nevertheless must assent to them, at least while I perceive them clearly; I remember that even before now, when I adhered very closely to the objects of the senses, I always took this type of truth to be the most certain of every truth that I evidently knew regarding figures, numbers, or other things pertaining to arithmetic, geometry or, in genera to pure and abstract mathematics.

But if, from the mere fact that I can bring forth from my thought the idea of something, it follows that all that I clearly and distinctly perceive to pertain to something really does pertain to it, then is this not an argument by which to prove the existence of God? Certainly I discover within me an idea of God, that is, of a supremely perfect being, no less than the idea of some figure or number. And I understand clearly and distinctly that it pertains to his nature that he always exists, no less than whatever has been demonstrated about some figure or number also pertains to the nature of this figure or number. Thus, even if everything that I have meditated upon during these last few days were not true, I ought to be at least as certain of the existence of God as I have hitherto been about the truths of mathematics.

66

Nevertheless, this point is not wholly obvious at first glance, but has the appearance of a sophism. Since in all other matters I am accustomed to distinguishing existence from essence, I easily persuade myself that the essence of God can be separated off from his existence; thus God can be thought of as not existing. Be that as it may, it still becomes obvious to a very diligently attentive person that the existence of God can no more be separated from his essence than the essence of a triangle can be separated from the fact that its three internal angles equal two right angles, or the idea of a valley can be separated from the idea of a mountain. So it is no less repugnant to think of a God (that is, a supremely perfect being) lacking existence (that is, as lacking some perfection), than it is to think of a mountain lacking a valley.

But granted I could no more think of God as not existing than I can think of a mountain without a valley, still it does not follow that a mountain actually exists in the world. Thus, from the fact that I think of God as existing, it does not seem to follow that God exists; for my thought imposes no necessity on things. Just as one can imagine a winged horse, without there being a horse with wings, so in the same way perhaps I can attach existence to God, even though no such God exists.

But there is a sophism lurking here. From the fact that I am unable to think of a mountain except with a valley, it does not follow that a mountain or a valley exists anywhere, but only that, whether they exist or not, a mountain and a valley cannot be separated from one another. But from the fact that I cannot think of God except as existing, it follows that existence is inseparable from God; for this reason he truly exists. Not because my thought brings this situation about, or imposes any necessity on anything; but because the necessity of this thing, namely of the existence of God, forces me to entertain this thought; for I am not free to think of God without existence (that is, the supremely perfect being apart from the supreme perfection), as I am free to imagine a horse with or without wings.

67

Further, it should not be said here that, having asserted that he has all perfections, I needed to assert that God exists—since existence is one of these perfections—but that my earlier assertion need not have been made. So, I need not believe that all four-sided figures are inscribed in a circle; but assuming that I do believe it, it would then be necessary for me to admit that a rhombus is inscribed in a circle—this is plainly false. Although it is not necessary that I happen upon any thought of God, nevertheless as often as I think of a being first and supreme—and bring forth the idea of God as if from the storehouse of my mind—I must of necessity ascribe all perfections to it, even though I do not at that time enumerate them all, nor take note of them one by one. This necessity plainly suffices so that afterwards, when I consider that existence is a perfection, I rightly conclude that a first and supreme being exists. In the same way, there is no necessity for me ever to imagine a triangle, but as often as I wish to consider a rectilinear figure having but three internal angles, I must ascribe to it those properties from which one rightly infers that the   68 three internal angles of this figure are not greater than two right angles, even though I do not then take notice of this fact. But when I inquire which figures might be inscribed in a circle, there is absolutely no need whatever for my believing that all four-sided figures are of this sort; nor even, for that matter, can I possibly imagine it, as long as I wish to admit only what I clearly and distinctly understand. Consequently, there is a great difference between false beliefs of this sort and true ideas inborn in me, the first and principal of which is the idea of God. For I plainly understand in many ways that it is not an invention dependent upon my thought, but an image of a true and immutable nature. First, because I cannot think of anything but God himself to whose essence belongs existence; next, because I cannot understand two or more Gods of this kind; because, having asserted that one God now exists, I plainly see that it is necessarily the case that he has existed from eternity and will endure forever; finally, because I perceive many other things in God, none of which I can remove or change.

But, indeed, whatever proof I use, it always comes down to the fact that the only things that fully convince me are those that I clearly and distinctly perceive. Although some of those things that I thus perceive are obvious to everyone—while others are discovered only by those who look more closely and inquire more diligently—nevertheless, after they have been discovered, they are considered no less certain than the others. For example, although it is not so readily apparent in the case of a right triangle that the square of the hypotenuse is equal to the square of the   69 other two sides as it is that the hypotenuse is opposite the largest angle; nonetheless, the former is no less believed—once it has been ascertained. However, I acknowledge nothing prior to, or easier than, what pertains to God, unless I am overwhelmed by prejudices or unless images of sensible things besiege my thought from all sides. For what in and of itself is more manifest than that a supreme being exists, or that God, to whose essence alone existence pertains, exists?

But, although I needed to pay close attention in order to perceive this point, nevertheless I not only am now as certain regarding this as I am regarding everything else that seems most certain, but also I observe that the certitude of other things so depends upon this point that without it I am unable ever to know anything perfectly.

Yet I am indeed of such a nature that, while I perceive something very clearly and distinctly, I cannot help but believe it to be true. And yet I am also of such a nature that I cannot keep the gaze of my mind always on the same thing in order to perceive it clearly; the memory of a judgment made earlier often comes to mind, when I no longer fully attend to the reasons for which I judged something. Thus, other reasons can be brought forward that, were I ignorant of God, would easily make me change my opinion; thus I would never have true and certain knowledge of anything, but I would have only vague and changeable opinions. Thus, for example, when I consider the nature of a triangle, it appears most evident to me—steeped as I am in the principles of geometry—that the three internal angles of the triangle are equal to two right angles; I must believe this to be true, as long as I attend to its demonstration. But as soon as I turn my power of discernment from the demonstration, even though I recall that I had looked at it most clearly, nevertheless, were I ignorant of God it can easily happen that I might doubt whether or not it is true. For I can convince myself that I have been so made by nature that I would occasionally be deceived about those things that I believe I quite evidently perceive, especially when I might recall that I have often taken many things to be true and certain, which later, upon bringing other considerations to bear, I have judged to be false.

But once I perceived that there is a God—because at the same time I also understood that all other things depend on him—and that he is no deceiver, I then concluded that everything that I clearly and distinctly perceive is necessarily true. Even if I no longer attend further to the reasons why I judged it to be true, provided that I remember that I did clearly and distinctly observe it, no contrary reason can be brought to bear which might force me to doubt it; rather, I have a true and certain knowledge of this matter. Not just this one either, but also everything else that I recall having once demonstrated, as in geometry, and so on. For what objections can now be raised to me? Have I been made such that I am often deceived? But now I know that I cannot be deceived in those matters that I clearly understand. On other occasions have I not taken many things to be true and certain, but that afterward I recognized as false? However, I did not clearly and distinctly perceive any of them, but—ignorant of this rule of truth—I believe them for other reasons, that I later discovered were less valid. What therefore will be said? Might I perhaps not be dreaming (as I recently objected to myself) or might everything about which I am now thinking be no more true than what passes before the mind of a person sleeping? Be that as it may, it changes nothing; for certainly, although I might be asleep, if something is evident to my intellect, it is wholly true.

71

And thus I plainly see that the certainty and truth of every science depends upon the knowledge of the true God, to such an extent that, before I had known him, I could not know anything perfectly about any other thing. But now it is possible for me to know certainly and fully countless things—both about God and other intellectual matters, as well as about all corporeal nature, which is the object of pure mathematics.

Meditation Six: Concerning the Existence of Material Things, and the Real Distinction of the Mind from the Body

It remains for me to examine whether material things exist. Indeed I now know that they can exist, at least insofar as they are the object of pure mathematics, because I clearly and distinctly perceive them. For no doubt God is capable of bringing about everything that I am thus capable of perceiving. And I have never judged that God was incapable of something, except when it was incompatible with being perceived by me distinctly. Moreover, from the faculty of imagination, which I observe that I use while I am engaged in dealing with these material things, it seems to follow that they exist; to someone paying very close attention to what imagination is, it appears to be nothing else but an application of the knowing faculty to a body intimately present to it—hence, a body that exists.

72

And to make this very clear, I first examine the difference between imagination and pure intellection. So, for example, when I imagine a triangle, I not only understand that it is a figure bounded by three lines, but at the same time I also intuit by my powers of discernment these three lines as present—this is what I call "imagining." But if I want to think about a chiliagon, I certainly understand just as well that it is a figure consisting of a thousand sides, and that a triangle is a figure consisting of three lines; but I do not imagine those thousand sides in the same way, that is, I do not intuit them as being present. Albeit that when

I think of a chiliagon I may perchance represent to myself some figure confusedly—because whenever I think about something corporeal, I always, out of force of habit, imagine something—nevertheless it is evident that it is not a chiliagon. This is so because it is not really different from the figure I would represent to myself if I were to think of a myriagon or any other figure with a large number of sides. Nor is imagination of any help in knowing the properties that differentiate the chiliagon from other polygons. But if it is a question of a pentagon, I surely can understand its form, just as was the case with the chiliagon, without the help of my imagination. But I can also imagine it, that is, by applying the powers of discernment both to its five sides and, at the same time, to the area bounded by those sides; clearly I am aware at this point that I need a peculiar sort of effort on the part of the mind in order to imagine, one that I do not employ in order to understand. This new effort on the part of the mind clearly shows the difference between imagination and pure intellection.

Besides, I believe that this power of imagining that is in me, insofar as it differs from the power of understanding, is not a necessary element of my essence, that is, of the essence of my mind; for although I might lack this power, nonetheless I would undoubtedly remain the same person as I am now. Thus it seems to follow that the power of imagining depends upon something different from me. And I readily understand that were a body to exist to which a mind is so joined that it might direct itself to look at it anytime it wishes, it could happen that by means of this body I intuit corporeal things. The result would be that this mode of thinking differs from pure intellection only in the fact that the mind, when it understands, in a sense turns itself toward itself and gazes upon one of the ideas that are in it. But when it imagines, it turns itself toward the body, and intuits something in the body similar to an idea either understood by the mind or perceived by sense. I say I easily understand that the imagination can function in this way, provided a body does exist. But this is only a probability; although I may investigate everything carefully, nevertheless, I do not yet see how, from the distinct idea of corporeal nature that I find in my imagination, I can draw up an argument that necessarily concludes that some body exists.

But I am in the habit of imagining many other things—over and above the corporeal nature that is the object of pure mathematics—such as colors, sounds, tastes, pain, and so on, but not so distinctly. Inasmuch as I perceive these things better with those senses from which, with the aid of the memory, they seem to have come to the imagination, the same trouble should also be taken concerning the senses, so that I might deal with them more appropriately. It must be seen whether I can obtain any certain argument for the existence of corporeal things from those things that are perceived by the way of thinking that I call "sense."

First I will repeat to myself here what those things were that I believed to be true because I had perceived them by means of the senses and what the grounds were for so believing. Next I will assess the reasons why I called them into doubt. Finally, I will consider what I must now believe concerning these things.

So, first, I sensed that I have a head, hands, feet, and other members that composed this body; I viewed it as a part of me, or perhaps even as the whole of me. I sensed that this body is frequently amid many other bodies, and that these bodies can affect my body in pleasant and unpleasant ways; I gauged what was pleasant by a certain sense of pleasure, and what was unpleasant by a sense of pain. In addition to pain and pleasure, I also sensed in me hunger, thirst, and other such appetites, as well as certain corporeal tendencies to mirth, sadness, anger, and other such feelings. But, as for things external to me—besides the extension, shapes, and motions of bodies—I also sensed their roughness, heat, and other tactile qualities; I sensed, too, the light, colors, odors, tastes, and sounds from whose variety I distinguished the heavens, the earth, the seas, and the other bodies one from the other. Indeed, because of the ideas of all these qualities that presented themselves to my thought and that alone I properly and immediately sensed, it was not wholly without reason that I believed that I sensed things clearly different from my thought, namely, the bodies from which these ideas might proceed. For I knew by experience that they come upon me without my consent, to the extent that, wish as I may, I cannot sense any object unless it be present to the organ of sense, and I cannot fail to sense it when it is present. Since the ideas perceived by sense are much more vivid and clear-cut, and even, in their own way, more "distinct," than any of those that I willingly and knowingly formed by meditation or those that I found impressed on my memory, it seems impossible that they come from myself. Therefore, it remained that they came from other things. Since I had no knowledge of such things except from these ideas themselves, I could not help entertaining the view that these things were similar to those ideas. Also, because I recalled that I had used my senses earlier than my reason, and I saw that the ideas that I myself constructed were not as clear-cut as those that I perceived by means of the senses, I saw that these former ideas were for the most part composed of parts of these latter ideas; I easily convinced myself that I plainly have no idea in the intellect that I did not have beforehand in the sense. Not without reason did I judge that this body, which by a special right I called "mine," belongs more to me than to any other thing, for I could never be separated from it in the same way as I could be from the rest. I sensed all appetites and feelings in it and for it. Finally, I have noticed pain and pleasurable excitement in its parts, but not in other bodies external to it. But why a certain sadness of spirit arises from one feeling of pain or another, and why a certain elation arises from a feeling of excitement, or why some sort of twitching of the stomach, which I call hunger, should warn me to take in nourishment, or why dryness of throat should warn me to take something to drink, and so on—for all these I plainly had no explanation other than that I have been taught so by nature. For there is clearly no affinity, at least none I am aware of, between the twitching of the stomach and the will to take in nourishment, or between the sense of something that causes pain and the thought of the sadness arising from this sense. But nature seems to have taught me everything else that I judged concerning the objects of the senses, because I convinced myself—even before having spent time on any of the arguments that might prove it—that things were this way.

But afterwards many experiences gradually caused all faith that I had in the senses to totter; occasionally towers, which had seemed round from afar, appeared square at close quarters; very large statues, standing on their pinnacles, did not seem large to someone looking at them from ground level; in countless other such things I detected that judgments of the external senses deceived me—not just the external senses, but also the internal senses. What can be more intimate than pain? But I had once heard it said by people whose leg or arm had been amputated that it seems to them that they occasionally sense pain in the very limb that they lacked. Therefore, even in me, it did not seem to be clearly certain that some part of my body was causing me pain, although I did sense pain in it. To these causes of doubt I recently added two quite general ones: first, I believed I never sensed anything while I was awake that I could not believe I also sometimes perceive while asleep. Since I do not believe that what I seem to sense in my dreams comes to me from things external to me, I did not see any reason why I should have these beliefs about things that I seem to sense while I am awake. The second cause of doubt was that, since I was ignorant of the cause of my coming into being (or at least pretended that I was ignorant of it), I saw that nothing prevented my having been so constituted by nature that I should be deceived about even what appeared to me most true. As to the arguments by which I formerly convinced myself of the truth of sensible things, I found no difficulty in responding to them. Since I seemed driven by nature toward many things opposed to reason, I did not think what was taught by nature deserved much credence. Although the perceptions of the senses did not depend on my will, I did not think that we must therefore conclude that they came from things external to me, because perhaps there is in me some faculty, as yet unknown to me, that produces these perceptions.

However, now, after having begun to know better the cause of my coming to be, I believe that I must not rashly admit everything that I seem to derive from the senses. But, then, neither should I call everything into doubt.

First, because I know that all the things that I clearly and distinctly understand can be made by God exactly as I understand them, it is enough that I can clearly and distinctly understand one thing without the other in order for me to be certain that the one thing is different from the other, because at least God can establish them separately. The question of the power by which this takes place is not relevant to their being thought to be different. For this reason, from the fact that I know that I exist, and that meanwhile I judge that nothing else clearly belongs to my nature or essence except that I am a thing that thinks, I rightly conclude that my essence consists in this alone: that I am only a thing that thinks. Although perhaps (or rather, as I shall soon say, to be sure) I have a body that is very closely joined to me, nevertheless, because on the one hand I have a clear and distinct idea of myself—insofar as I am a thing that thinks and not an extended thing—and because on the other hand I have a distinct idea of a body—insofar as it is merely an extended thing, and not a thing that thinks—it is therefore certain that I am truly distinct from my body, and that I can exist without it.

Moreover, I find in myself faculties endowed with certain special modes of thinking—namely the faculties of imagining and sensing—without which I can clearly and distinctly understand myself in my entirety, but not vice versa: I cannot understand them clearly and distinctly without me, that is, without the knowing substance to which they are attached. For in their formal concept they include an act of understanding; thus I perceive that they are distinguished from me just as modes are to be distinguished from the thing of which they are modes. I also recognize certain other faculties—like those of moving from one place to another, of taking on various shapes, and so on—that surely no more can be understood without the substance to which they are attached than those preceding faculties; for that reason they cannot exist without the substance to which they are attached. But it is clear that these faculties, if in fact they exist, must be attached to corporeal or extended substances, but not to a knowing substance, because extension—but certainly not understanding—is contained in a clear and distinct concept of them. But now there surely is in me a passive faculty of sensing, that is, of receiving and knowing the ideas of sensible things; but I cannot use it unless there also exists, either in me or in something else, a certain active faculty of producing or bringing about these ideas. This faculty surely cannot be in me, since it clearly presupposes no intellection, and these ideas are produced without my cooperation and often against my will. Because this faculty is in a substance other than myself, in which ought to be contained—formally or eminently— all the reality that is objectively in the ideas produced by this faculty (as I have just now taken notice), it

79

thus remains that either this substance is a body (or corporeal nature) in which is contained formally all that is contained in ideas objectively or it is God—or some other creature more noble than a body—in which it is all contained eminently. But, since God is not a deceiver, it is absolutely clear that he sends me these ideas neither directly and immediately—nor even through the mediation of any creature, in which the objective reality of these ideas is contained not formally but only eminently. Since he plainly gave me no faculty for making this discrimination—rather, he gave me a great inclination to believe that these ideas proceeded from corporeal things—I fail to see why God cannot be understood to be a deceiver, if they proceeded from a source other than corporeal things. For this reason, corporeal things exist. Be that as it may, perhaps not all bodies exist exactly as I grasp them by sense, because this grasp by the senses is in many cases very obscure and confused. But at least everything is in these bodies that I clearly and distinctly understand—that is, everything, considered in a general sense, that is encompassed in the object of pure mathematics.

But as to how this point relates to the other remaining matters that are either merely particular—as, for example, that the sun is of such and such a size or shape, and so on—or less clearly understood—as, for example, light, sound, pain, and so on—although they are very doubtful and uncertain, still, because God is not a deceiver, and no falsity can be found in my opinions, unless there is also in me a faculty given me by God for the purpose of rectifying this falsity, these features provide me with a certain hope of reaching the truth in them. And plainly it cannot be doubted that whatever I am taught by nature has some truth to it; for by "nature," taken generally, I understand only God himself or the co-ordination, instituted by God, of created things. I understand nothing else by my nature in particular than the totality of all the things bestowed on me by God.

There is nothing that this nature teaches me in a more clear-cut way than that I have a body that is ill-disposed when I feel pain, that it needs food and drink when I suffer hunger or thirst, and so on. Therefore, I ought not to doubt that there is some truth in this.

By means of these feelings of pain, hunger, thirst and so on, nature also teaches that I am present to my body not merely in the way a seaman is present to his ship, but that I am tightly joined and, so to speak, mingled together with it, so much so that I make up one single thing with it. For otherwise, when the body is wounded, I, who am nothing but a thing that thinks, would not then sense the pain. Rather, I would perceive the wound by means of the pure intellect, just as a seaman perceives by means of sight whether anything in the ship is broken. When the body lacks food or drink, I would understand this in a clear-cut fashion; I would not have confused feelings of hunger and thirst. For certainly these feelings of thirst, hunger, pain, and so on are nothing but confused modes of thinking arising from the union and, as it were, the mingling of the mind with the body.

Moreover, I am also taught by nature that many other bodies exist around my body; some of them are to be pursued, and others are to be avoided. And to be sure, from the fact that I sense widely different colors, sounds, odors, tastes, heat, roughness, and so on, I rightly conclude that in the bodies from which these different perceptions of the senses proceed, there are differences corresponding to the different perceptions—although perhaps the former are not similar to the latter. But from the fact that some of these perceptions are pleasant and others unpleasant, it is plainly certain that my body—or rather my whole self, insofar as I am composed of a body and a mind—can be affected by various benefits and harms from the surrounding bodies.

But I have accepted many other things, and, although I seem to have 82 been taught them by nature, still it was not really nature that taught them to me, but a certain habit of making unconsidered judgments. And thus it could easily happen that they are false, as, for example, the belief that any space where there is nothing that moves my senses is empty; or, for example, the belief that in a hot body there is something plainly similar to the idea of heat which is in me; or that in a white or green body there is the same whiteness or greeness that I sense; or in a bitter or sweet thing the same taste, and so on; or the belief that stars and towers, and any other distant bodies have only the same size and shape that they present to the senses—and other examples of the same sort. But lest I not perceive distinctly enough something about this matter, I ought to define more carefully what I properly understand when I say that I am "taught something by nature." For I am using "nature" here in a stricter sense than the totality of everything bestowed on me by God. For in this totality there are contained many things that pertain only to my mind, as, for example, that I perceive that what has been done cannot be undone, and everything else that is known by the light of nature. At the moment the discussion does not center on these matters. There are also many things that pertain only to the body, as, for example, that it tends downward, and so on. I am not dealing with these either, but only with what has been bestowed on me by God, insofar as I am composed of mind and body. Therefore it is nature, thus understood, that teaches me to flee what brings a sense of pain and to pursue what brings a sense of pleasure, and the like. But it does not appear that nature, so conceived, teaches that we conclude from these perceptions of the senses anything in addition to this regarding things external to us unless there previously be an inquiry by the intellect; for it pertains to the mind alone, and not to the

83 composite, to know the truth in these matters. Thus, although a star affects my eye no more than the flame from a small torch, still there is no real or positive tendency in my eye toward believing that the star is any bigger than the flame; rather, ever since my youth, I have made this judgment without reason. Although I feel heat upon drawing closer to the fire, and I feel pain upon drawing even closer to it, there is indeed no argument that convinces me that there is something in the fire that is similar either to the heat or to the pain, but only that there is something in the fire that causes in us these feelings of heat or pain. Although there be nothing in a given space that moves the sense, it does not therefore follow that there is no body in it. I use the perceptions of the senses that properly have been given by nature only for the purpose of signifying to the mind what is agreeable and disagreeable to the composite, of which the mind is a part; within those limits these perceptions are sufficiently clear and distinct. However, I see that I have been in the habit of subverting the order of nature in these and many other matters, because I use the perceptions of the senses as certain rules for immediately discerning what the essence is of the bodies external to us; yet, in respect to this essence, these perceptions still show me nothing but obscurity and confusion.

I have already examined in sufficient detail how it could happen that my judgments are false, the goodness of God notwithstanding. But a new difficulty now comes on the scene concerning those very things that are shown to me by nature as things to be either sought or avoided, as well as concerning the internal senses in which I seem to have detected errors: for example, when a person, deluded by the pleasant taste

84 of food, ingests a poison hidden inside it. But in this case he is impelled by nature only toward desiring the thing in which the pleasant taste is located, but not toward the poison, of which he obviously is unaware. Nothing else can be concluded here except that this nature is not all-knowing. This is not remarkable, since man is a limited being; thus only limited perfection is appropriate to man.

But we often err even in those things to which nature impels us; for example, when those who are ill desire food or drink that will soon be injurious to them. Perhaps it could have been said here that they erred because their nature was corrupt. But this does not remove our difficulty, because a sickly man is no less a creature of God than a healthy one; for that reason it does not seem any less repugnant that the sickly man got a deceiving nature from God. And just as a clock made of wheels and counter-weights follows all the laws of nature no less closely when it has been badly constructed and does not tell time accurately than when it satisfies on all scores the wishes of its maker, just so, if I should consider the body of a man—insofar as it is a kind of mechanism composed of and outfitted with bones, nerves, muscles, veins, blood and skin—even if no mind existed in it, the man's body would still have all the same motions that are in it now except for those motions that proceed either from a command of the will or, consequently, from the mind. I readily recognize that it would be natural for this body, were it, say, suffering from dropsy, to suffer dryness of the throat, which commonly brings a feeling of thirst to the mind, and thus too its nerves and other parts are so disposed by the mind to take a drink with the result that the sickness is increased. It would be no more natural for this body, when there is no such infirmity in it and it is moved by the same dryness, to drink something useful to it. And, from the point of view of the intended purpose of the watch, I could say that it turns away from its nature when it does not tell the right time. Similarly, considering the mechanism of the human body as equipped for the motions that typically occur in it, I might think that it too turns away from its nature, if its throat were dry, when taking a drink would not be beneficial to its continued existence. Nevertheless, I realize well enough that this latter usage of the term "nature" differs greatly from the former. For this latter "nature" is only an arbitrary denomination, extrinsic to the things on which it is predicated and dependent upon my thought, because it compares a man in poor health and a poorly constructed clock with the idea of a man in good health and a well-made clock. But by "nature" taken in the former sense, I understand something that really is in things, and thus it is not without some truth.

When it is said, in the case of the dropsical body, that its "nature" is corrupt—from the fact that this body has a parched throat, and yet does not need a drink—it certainly is only an extrinsic denomination of nature. But be that as it may, in the case of the composite, that is, of a mind joined to such a body, it is not a pure denomination, but a true error of nature that this body should thirst when a drink would be harmful to it. Therefore it remains here to inquire how the goodness of God does not stand in the way of "nature," thus considered, being deceptive.

Now, first, I realize at this point that there is a great difference between a mind and a body, because the body, by its very nature, is something divisible, whereas the mind is plainly indivisible. Obviously, when I consider the mind, that is, myself insofar as I am only a thing that thinks, I cannot distinguish any parts in me; rather, I take myself to be one complete thing. Although the whole mind seems to be united to the whole body, nevertheless, were a foot or an arm or any other bodily part amputated, I know that nothing would be taken away from the mind; nor can the faculties of willing, sensing, understanding, and so on be called its "parts," because it is one and the same mind that wills, senses, and understands. On the other hand, no corporeal or extended thing can be thought by me that I did not easily in thought divide into parts; in this way I know that it is divisible. If I did not yet know it from any other source, this consideration alone would suffice to teach me that the mind is wholly different from the body.

Next, I observe that my mind is not immediately affected by all the parts of the body, but merely by the brain, or perhaps even by just one small part of the brain—namely, by that part in which the "common sense" is said to be found. As often as it is disposed in the same manner, it presents the same thing to the mind, although the other parts of the body can meanwhile orient themselves now this way, now that way, as countless experiments show—none of which need be reviewed here.

I also notice that the nature of the body is such that none of its parts can be moved by another part a short distance away, unless it is also moved in the same direction by any of the parts that stand between them, even though this more distant part does nothing. For example, in the cord ABCD, if the final part D is pulled, the first part A would be moved in exactly the same direction as it could be moved if one of the intermediate parts, B or C, were pulled and the last part D remained motionless. Just so, when I sense pain in the foot, physics teaches me that this feeling took place because of nerves scattered throughout the foot. These nerves, like cords, are extended from that point all the way to the brain; when they are pulled in the foot, they also pull on the inner parts of the brain to which they are stretched, and produce a certain motion in these parts of the brain. This motion has been constituted by nature so as to affect the mind with a feeling of pain, as if it existed in the foot. But because these nerves need to pass through the tibia, thigh, loins, back, and neck, with the result that they extend from the foot to the brain, it can happen that the part that is in the foot is not stretched; rather, one of the intermediate parts is thus stretched, and obviously the same movement will occur in the brain that happens when the foot was badly affected. The necessary result is that the mind feels the same pain. And we must believe the same regarding any other sense.

Finally, I observe that, since each of the motions occurring in that part of the brain that immediately affects the mind occasions only one sensation in it, there is no better way to think about this than that it occasions the sensation that, of all that could be occasioned by it, is most especially and most often conducive to the maintenance of a healthy man. Moreover, experience shows that such are all the senses bestowed on us by nature; therefore, clearly nothing is to be found in them that does not bear witness to God's power and goodness. Thus, for example, when the nerves in the foot are violently and unusually agitated, their motion, which extends through the marrow of the spine to the inner reaches of the brain, gives the mind at that point a sign to feel something—namely, the pain as if existing in the foot. This pain provokes it to do its utmost to move away from the cause, since it is harmful to the foot. But the nature of man could have been so constituted by God that this same motion in the brain might have displayed something else to the mind: either the motion itself as it is in the brain, or as it is in the foot, or in some place in between—or somewhere else entirely different. But nothing else serves so well the maintenance of the body. Similarly, when we need a drink, a certain dryness arises in the throat that moves its nerves, and, by means of them, the inner recesses of the brain. This motion affects the mind with a feeling of thirst, because in this situation nothing is more useful for us to know than that we need a drink to sustain our health; the same holds for the other matters.

From these considerations it is totally clear that, notwithstanding the immense goodness of God, the nature of man—insofar as it is composed of mind and body, cannot help but sometimes be deceived. For if some cause, not in the foot but in some other part through which the nerves are stretched from the foot to the brain—or perhaps even in the brain itself—were to produce the same motion that would normally be produced by a badly affected foot, then the pain will be felt as if it were in the foot, and the senses will naturally be deceived, because it is reasonable that the motion should always show the pain to the mind as something belonging to the foot rather than to some other part, since an identical motion in the brain can bring about only the identical effect and this motion more frequently is wont to arise from a cause that harms the foot than from something existing elsewhere. And if the dryness of the throat does not, as is the custom, arise from the fact that drink aids in the health of the body, but from a contrary cause—as happens in the case of the person with dropsy—then it is fâr better that it should deceive, than if, on the contrary, it were always deceptive when the body is well constituted. The same goes for the other cases.

This consideration is most helpful, not only for noticing all the errors to which my nature is liable, but also for easily being able to correct or avoid them. To be sure, I know that every sense more frequently indicates what is true than what is false regarding those things that concern the advantage of the body, and I can almost always use more than one sense in order to examine the same thing. Furthermore, I can use memory, which connects present things with preceding ones, plus the intellect, which now has examined all the causes of error. I should no longer fear lest those things that are daily shown me by the senses, are false; rather, the hyperbolic doubts of the last few days ought to be rejected as worthy of derision—especially the principal doubt regarding sleep, which I did not distinguish from being awake. For I now notice that a very great difference exists between these two; dreams are never joined with all the other actions of life by the memory, as is the case with those actions that occur when one is awake. For surely, if someone, while I am awake, suddenly appears to me, and then immediately disappears, as happens in dreams, so that I see neither where he came from or where he went, it is not without reason that I would judge him to be a ghost or a phantom conjured up in my brain, rather than a true man. But when these things happen, regarding which I notice distinctly where they come from, where they are now, and when they come to me, and I connect the perception of them without any interruption with the rest of my life, obviously I am certain that these perceptions have occurred not in sleep but in a waking state. Nor ought I to have even a little doubt regarding the truth of these things, if, having mustered all the senses, memory, and intellect in order to examine them, nothing is announced to me by one of these sources that conflicts with the others. For from the fact that God is no deceiver, it follows that I am in no way deceived in these matters. But because the need to get things done does not always give us the leisure time for such a careful inquiry, one must believe that the life of man is vulnerable to errors regarding particular things, and we must acknowledge the infirmity of our nature.